MATH MASTERS

Everyday
Mathematics®

The University of Chicago School Mathematics Project

Mc
Graw
Hill
Education

The University of Chicago School Mathematics Project

Max Bell, Director, *Everyday Mathematics* First Edition; James McBride, Director, *Everyday Mathematics* Second Edition; Andy Isaacs, Director, *Everyday Mathematics* Third, CCSS, and Fourth Editions; Amy Dillard, Associate Director, *Everyday Mathematics* Third Edition; Rachel Malpass McCall, Associate Director, *Everyday Mathematics* CCSS and Fourth Editions; Mary Ellen Dairyko, Associate Director, *Everyday Mathematics* Fourth Edition

Authors
Max Bell, John Bretzlauf, Amy Dillard, Robert Hartfield, Andy Isaacs, Rebecca Williams Maxcy, James McBride, Kathleen Pitvorec, Peter Saecker, Robert Balfanz*, William Carroll*, Sheil Sconiers*

*First Edition Only

Fourth Edition Grade 4 Team Leader
Rebecca Williams Maxcy

Writers
Meg Schleppenbach Bates, Randee Blair, Kristin Fitzgerald, Carla LaRochelle, Sara A. Snodgrass

Open Response Team
Catherine R. Kelso, Leader; Judith S. Zawojewski, Andy Carter, John Benson

Differentiation Team
Ava Belisle-Chatterjee, Leader; Jean Capper, Martin Gartzman, Barbara Molina

Digital Development Team
Carla Agard-Strickland, Leader; John Benson, Gregory Berns-Leone, Juan Camilo Acevedo

Virtual Learning Community
Meg Schleppenbach Bates, Cheryl G. Moran, Margaret Sharkey

Technical Art
Diana Barrie, Senior Artist; Cherry Inthalangsy

UCSMP Editorial
Don Reneau, Senior Editor; Elizabeth Olin, Kristen Pasmore, Lucas Whalen

Field Test Coordination
Denise A. Porter

Field Test Teachers
Kindra Arwood, Tiffany N. Harper, Brian A. Herman, Tonya Howell, Amy Jacobs, Amy Jarrett-Clancy, Kari Lehman, Stephanie Rogers, Jenna Rose Ryan, Rachel Schrader, JoAnn Tennenbaum, Robin Zogby

Digital Field Test Teachers
Colleen Girard, Michelle Kutanovski, Gina Cipriani, Retonyar Ringold, Catherine Rollings, Julia Schacht, Christine Molina-Rebecca, Monica Diaz de Leon, Tiffany Barnes, Andrea Bonanno-Lersch, Debra Fields, Kellie Johnson, Elyse D'Andrea, Katie Fielden, Jamie Henry, Jill Parisi, Lauren Wolkhamer, Kenecia Moore, Julie Spaite, Sue White, Damaris Miles, Kelly Fitzgerald

Contributors
William Baker, John Benson, Jeanne Di Domenico, Jim Flanders, Lila Schwartz, Penny Williams

Center for Elementary Mathematics and Science Education Administration
Martin Gartzman, Executive Director; Meri B. Fohran, Jose J. Fragoso, Jr., Regina Littleton, Laurie K. Thrasher

External Reviewers

The *Everyday Mathematics* authors gratefully acknowledge the work of the many scholars and teachers who reviewed plans for this edition. All decisions regarding the content and pedagogy of *Everyday Mathematics* were made by the authors and do not necessarily reflect the views of those listed below.

Elizabeth Babcock, California Academy of Sciences; Arthur J. Baroody, University of Illinois at Urbana-Champaign and University of Denver; Dawn Berk, University of Delaware; Diane J. Briars, Pittsburgh, Pennsylvania; Kathryn B. Chval, University of Missouri–Columbia; Kathleen Cramer, University of Minnesota; Ethan Danahy, Tufts University; Tom de Boor, Grunwald Associates; Louis V. DiBello, University of Illinois at Chicago; Corey Drake, Michigan State University; David Foster, Silicon Valley Mathematics Initiative; Funda Gönüleş, Michigan State University; M. Kathleen Heid, Pennsylvania State University; Natalie Jakucyn, Glenbrook South High School, Glenview, IL; Richard G. Kron, University of Chicago; Richard Lehrer, Vanderbilt University; Susan C. Levine, University of Chicago; Lorraine M. Males, University of Nebraska-Lincoln; Dr. George Mehler, Temple University and Central Bucks School District, Pennsylvania; Kenny Huy Nguyen, North Carolina State University; Mark Oreglia, University of Chicago; Sandra Overcash, Virginia Beach City Public Schools, Virginia; Raedy M. Ping, University of Chicago; Kevin L. Polk, Aveniros LLC; Sarah R. Powell, University of Texas at Austin; Janine T. Remillard, University of Pennsylvania; John P. Smith III, Michigan State University; Mary Kay Stein, University of Pittsburgh; Dale Truding, Arlington Heights District 25, Arlington Heights, Illinois; Judith S. Zawojewski, Illinois Institute of Technology

Note
Many people have contributed to the creation of *Everyday Mathematics*. Visit http://everydaymath.uchicago.edu/authors/ for biographical sketches of *Everyday Mathematics* 4 staff and copyright pages from earlier editions.

www.everydaymath.com

Send all inquiries to:
McGraw-Hill Education
8787 Orion Place
Columbus, OH 43240

ISBN: 978-0-02-137658-2
MHID: 0-02-137658-1

Printed in the United States of America.

1 2 3 4 5 6 7 8 9 RHR 20 19 18 17 16 15

Contents

Teaching Masters and Home Link Masters

Unit 1

Unit 2

Unit 3

Unit 4

Unit 5

Contents **vii**

Teaching Aid Masters

Game Masters

Teaching Masters and Home Link Masters

7-Digit Place-Value Chart

Millions	Hundred-Thousands	Ten-Thousands	Thousands	Hundreds	Tens	Ones

Place Value in Whole Numbers

Family Note In this lesson your child explored the relationships between place values in numbers. Have your child read each number below. Examine the digit 6 in each number.

Hundred-Thousands	Ten-Thousands	Thousands	Hundreds	Tens	Ones
600,000	60,000	6,000	600	60	6

When the digit 6 moves left one place, its value becomes 10 times as large as it was in the previous place. For example, 60 is 10 times as large as 6, and 600 is 10 times as large as 60.

(1) **a.** The 8 in 203,810 is worth _____.   **b.** The 6 in 56,143 is worth _____.

SRB 78-79

c. The 7 in 573,090 is worth _____.   **d.** The 1 in 140,007 is worth _____.

(2) How does the value of the digit 4 in 489 differ from the value of the digit 4 in 5,741?

_____

_____

(3) **a.** The value of 8 in 56,982 is _____ times as large as the value of 8 in 156,408.

b. The value of 8 in 800 is _____ times as large as the value of 8 in 80.

c. The value of 9 in 4,934 is _____ times as large as the value of 9 in 1,290.

(4) **a.** Write the number that has . . .
7 in the thousands place
6 in the ten-thousands place
5 in the hundreds place
8 in the ones place
3 in the tens place

_____ _____ , _____ _____ _____

b. On the back of this page, write this number in words.

Practice

(5) $9 + 8 =$ _____   (6) $7 + 8 =$ _____   (7) $30 + 80 =$ _____

(8) _____ $= 50 + 40$ (9) _____ $= 17 + 94$ (10) $158 + 93 =$ _____

Introduction to *Fourth Grade Everyday Mathematics*

Welcome to *Fourth Grade Everyday Mathematics*, part of an elementary school mathematics curriculum developed by the University of Chicago School Mathematics Project (UCSMP).

Fourth Grade Everyday Mathematics emphasizes the following content:

Operations and Algebraic Thinking Investigating methods for solving problems involving mathematics in everyday situations; solving multistep problems involving the four operations; using estimation to check the reasonableness of answers; exploring properties of numbers such as multiples, factor pairs, prime and composite; and designing, exploring, and using geometric and number patterns.

Numbers and Operations in Base Ten Reading, writing, comparing, and ordering whole numbers; exploring addition, subtraction, multiplication, and division methods; inventing individual procedures and methods.

Number and Operations—Fractions Developing an understanding of fraction equivalence; exploring addition and subtraction of fractions with like denominators and multiplication of fractions by whole numbers; and reading, writing, comparing, and ordering fractions and decimals.

Measurement and Data Exploring metric and U.S. customary measurement systems and converting from larger units to smaller units within a single system; applying formulas to find the perimeters and areas of rectangles; developing an understanding of angles and angle measurement; and representing and interpreting data on line plots.

Geometry Drawing and identifying geometric properties and identifying these properties in polygons; recognizing and drawing a line of symmetry; identifying symmetric figures.

Everyday Mathematics provides you with many opportunities to monitor your child's progress and participate in your child's experience of mathematics. Throughout the year you will receive Family Letters to keep you informed of the mathematical content your child will be studying in each unit. Each letter includes a vocabulary list, suggested Do-Anytime Activities for you and your child, and an answer guide to selected Home Link (homework) activities. You will enjoy seeing your child's confidence and comprehension soar as he or she connects mathematics to everyday life.

This unit reviews and extends mathematical content developed in *Third Grade Everyday Mathematics.* In Unit 1, students will explore the following concepts:

Place Value in Whole Numbers Students review place-value concepts and explore numbers in the ten-thousands and hundred-thousands. They will read, write, compare, and order these numbers. Students will also use population data from U.S. cities to practice rounding and comparison techniques.

Computation Students practice mental and paper-and-pencil methods of computation, as well as using a calculator. They will decide which tool is most appropriate for solving a particular problem.

Students explore a new strategy for adding and subtracting multidigit whole numbers and compare different methods. They will realize that often the same result may be obtained in multiple ways.

Students use estimation to assess the reasonableness of answers as they work with multistep number stories using a letter for the unknown.

Measurement and Data Students review the concept of perimeter and then develop and apply formulas for finding the perimeters of rectangles.

Students convert between customary units of length (yards, feet, inches) and solve number stories involving conversions.

Geometry Students examine definitions and properties of 2-dimensional shapes and the relationships among them.

Please keep this Family Letter for reference as your child works through Unit 1.

Vocabulary
Important terms in Unit 1:

acute angle An angle with a measure greater than 0° and less than 90°.

acute angle

angle A figure that is formed by two rays or two line segments with a common endpoint.

base 10 Our number system in which each place in a number has a value 10 times as large as the place to its right and $\frac{1}{10}$ the place to its left.

digit One of the number symbols 0, 1, 2, 3, 4, 5, 6, 7, 8, and 9 in the standard base-10 system.

endpoint A point at the end of a line segment or ray.

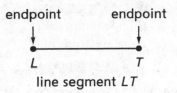

endpoint   endpoint

L   *T*

line segment *LT*

expanded form A way of writing a number as the sum of the values of each digit. For example, the expanded form of 356 is 300 + 50 + 6.

intersect To share a common point or points.

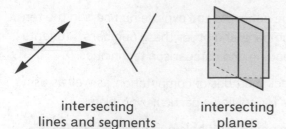

intersecting
lines and segments

intersecting
planes

line A straight path that extends infinitely in opposite directions.

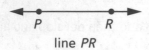

line *PR*

line segment A straight path joining two points, which are called endpoints.

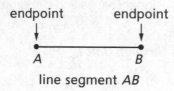

line segment *AB*

obtuse angle An angle that has a measure greater than 90° and less than 180°.

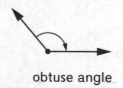

obtuse angle

parallel Lines, line segments, or rays in the same plane are parallel if they never cross or meet, no matter how far they are extended in either direction.

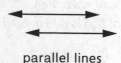

parallel lines

parallelogram A quadrilateral that has two pairs of parallel sides. Opposite sides of a parallelogram have equal lengths, and its opposite angles have the same measure.

perimeter The distance around the boundary of a 2-dimensional figure.

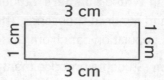

perimeter = 1 cm + 3 cm + 1 cm + 3 cm = 8 cm

perpendicular Crossing or meeting at right angles. Lines, rays, or line segments that cross or meet at right angles are perpendicular.

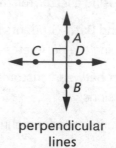

perpendicular
lines

place value The value given to a digit according to its position, or place, in a number. The chart on the next page shows the value of each digit in 24,815.

point An exact location in space. Lines have an infinite number of points on them.

ray A straight path that extends infinitely from an endpoint. A ray is named using the letter label of its endpoint followed by the letter label of another point on the ray.

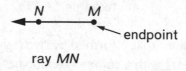

ray *MN*

right angle An angle that measures exactly 90°.

right triangle A triangle that contains a right angle.

right triangle

U.S. traditional addition A paper-and-pencil method for adding multidigit numbers in which the addends are stacked vertically with place values aligned and the digits in each column are added, working column by column from the right. The tens digit, if any, from a column sum is "carried" to the top of the next column to the left and is added with the digits in that column.

U.S. traditional subtraction A paper-and-pencil method for subtracting multidigit numbers. The minuend (number from which another is subtracted) and subtrahend (number being subtracted) are stacked vertically with place values aligned and the digits in each column are subtracted, working column by column from the right. If a digit in the subtrahend is larger than the corresponding digit in the minuend, a 10 is "borrowed" from the next column to the left.

vertex A point at which the rays of an angle or the sides of a geometric figure meet.

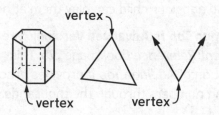

whole numbers The numbers 0, 1, 2, 3, 4, and so on.

Ten-Thousands	Thousands	Hundreds	Tens	Ones
2	4	8	1	5
The value of the 2 is 20,000.	The value of the 4 is 4,000.	The value of the 8 is 800.	The value of the 1 is 10.	The value of the 5 is 5.

Place-value chart

Do-Anytime Activities

To work with your child on concepts taught in this unit, try these activities:

1. Have your child locate big numbers in newspapers and other sources and ask him or her to read them to you. Or read the numbers and have your child write them down.

2. Help your child look up the populations and land areas of the state and city in which you live and compare them with the populations and areas of other states and cities.

3. Together, write five multidigit numbers in order from smallest to largest.

4. Model real-life uses of estimation for your child. For example, when you are shopping, round the cost of several items up to the nearest dollar and add to estimate their total cost.

5. Help your child discover everyday uses of geometry found in art, architecture, jewelry, toys, and so on.

Building Skills through Games

Throughout the school year, students will play mathematics games as a way to practice a variety of arithmetic skills. Playing games turns practice into a fun thinking activity. Games in this unit provide practice with place value, addition, and subtraction. They require very few materials, so you and your child can play them at home.

***Addition Top-It* (Advanced Version)** See *Student Reference Book,* page 275. This variation of *Addition Top-It* provides practice adding numbers through the thousands.

Fishing for Digits See *Student Reference Book,* page 259. This game provides practice identifying digits and their values, as well as adding and subtracting.

Number Top-It See *Student Reference Book,* page 269. This game provides practice working with place value through the hundred-thousands.

***Subtraction Top-It* (Advanced Version)** See *Student Reference Book,* page 275. This variation of *Subtraction Top-It* provides practice subtracting numbers through the thousands.

As You Help Your Child with Homework

As your child brings assignments home, you may want to go over the instructions together, clarifying them as necessary. The answers listed below will guide you through the Home Links for this unit.

Home Link 1-1
1. **a.** 800 **b.** 6,000 **c.** 70,000 **d.** 100,000
3. 10; 10; 10
4. 67,538 ; Sixty-seven thousand, five hundred thirty-eight

Home Link 1-2
3a. Uganda. Both have the same number of ten-thousands, but Uganda's area has 3 thousands and Laos's area has 1 thousand.

3b. 93,100 > 91,400

Home Link 1-3
1. Chicago Sky: 18,000
   Connecticut Sun: 10,000
   Indiana Fever: 18,000
   Los Angeles Sparks: 13,000
   Minnesota Lynx: 19,000
   Phoenix Mercury: 18,000
   Seattle Storm: 17,000
   Tulsa Shock: 18,000
   Washington Mystics: 20,000
3. Wyoming: 600,000
   Vermont: 600,000
   North Dakota: 700,000
   Alaska: 700,000
   South Dakota: 800,000

Home Link 1-4
1. Baseball
3. 3,000,000; 4,000,000; 4,000,000; 3,000,000; 2,000,000
5. 2,370,794 < 3,565,718

Home Link 1-5

1. **a.** No. Sample answer: I rounded the times to the tens place and added: 20 + 40 = 60 and 60 + 20 = 80.

    b. Sample answer: The numbers were all close to a multiple of 10. I just needed to know if they added up to more or less than 60.

2. **a.** No. Sample answer: I used close-but-easier numbers. I rounded 31 to 30 and 24 to 25. 30 + 25 = 55; 100 − 55 = 45.

    b. Sample answer: The numbers were all close to friendly numbers, so I decided to go with close-but-easier numbers.

Home Link 1-6

Estimates vary.

1. 150 pounds; 144 pounds; Sample answer: 700 − (176 + 250 + 130) = 144; Yes. Sample answer: My estimate was 150 pounds, which is close to my answer.

2. 300 pounds; 272 pounds; Sample answer: (491 − (175 + 180)) * 2 = 272; Yes; Sample answer: My estimate was 300 pounds, which is close to 272 pounds.

Home Link 1-7

Estimates vary.

1. 82; 40 + 50 = 90   **3.** 1,673; 800 + 900 = 1,700

5. 2,074; 300 + 1,800 = 2,100   **7.** 2,800; 3,000

Home Link 1-8

1. 100 balls

2. 730 balls; Sample answer: 7 * 100 balls in a carton + 3 * 10 balls in a box = 730 balls

3. Sample answer: The number of cartons is like the digit in the 100s place, and the number of boxes is like the digit in the 10s place.

Home Link 1-9

Estimates vary.

1. 47; 90 − 40 = 50   **3.** 319; 500 − 200 = 300

5. 795; 2,000 − 1,000 = 1,000   **7.** 1,034

Home Link 1-10

1. 12; 72; 96; 144   **3.** 27 feet

5. Four thousand, eight hundred fifty-seven

Home Link 1-11

3. **a.**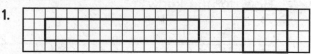

    b. No. A ray's endpoint must be listed first when naming a ray.

4. $\overline{WX}$ (or $\overline{XW}$) is parallel to $\overline{ST}$ (or $\overline{TS}$).

Home Link 1-12

1.

3. Both have right angles and perpendicular sides. They have a different number of sides and right angles.

4. **a.**   **b.** E   **c.** ∠FED

5.

Home Link 1-13

1. 30 feet   **2.** 42 inches   **3.** 116 feet

4. 108 inches   **5.** 6 feet

7. 900,000; 900,000

Comparing 6-Digit Numbers

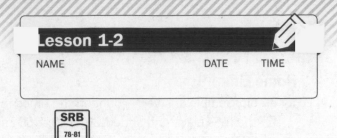

Write your 3 numbers on the lines below.

SRB
78-81

_____

_____

_____

① Write a number greater than your largest number. _____

② Write a number that falls between your two largest numbers. _____

③ Which of your numbers is closest to 500,000? _____

How do you know?

_____

④ Which of your numbers is closest to 100,000? _____

How do you know?

_____

⑤ Write your largest number in expanded form.

_____

⑥ Choose any two of your numbers and write a comparison number sentence
using <, >, or =.

_____

⑦ Compare your smallest number with your partner's smallest number.
Write a comparison number sentence using <, >, or =.

_____

⑧ Write your smallest number in expanded form.

_____

⑨ Write your smallest number in words.

_____

Country Sizes

This table shows the sizes of
10 countries measured in square miles.

Use a place-value tool to help you answer
the questions.

SRB 81

① Read the numbers to someone at home.

② Which is the largest country listed?

_____

The smallest? _____

Country	Area (in square miles)
Algeria	919,600
Colombia	439,700
Ethiopia	426,400
Egypt	386,700
Greece	50,900
Iran	636,400
Laos	91,400
Peru	494,200
Uganda	93,100

Source: worldatlas.com (All data rounded to nearest hundred.)

③ Compare the areas of Laos and Uganda.

   a. Which country has the larger area? _____ How do you know?

   _____

   _____

   b. Write a comparison number sentence. _____

④ Order the countries from largest
area to smallest area.

Country	Area (in square miles)

Practice

⑤ 140 − 60 = _____   ⑥ _____ = 57 − 39   ⑦ 115 − 86 = _____

Finding the Halfway Point

① For each number line, record the number that is halfway between the lower and higher numbers. Then plot a number that is *less* than the halfway number.

a.

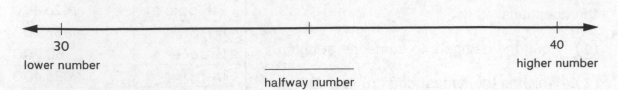

30
lower number

_____
halfway number

40
higher number

b.

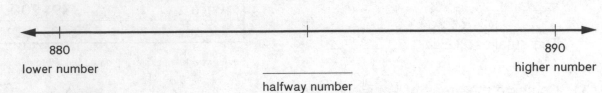

880
lower number

_____
halfway number

890
higher number

② For each number line, record the number that is halfway between the lower and higher numbers. Then plot a number that is *greater* than the halfway number.

a.

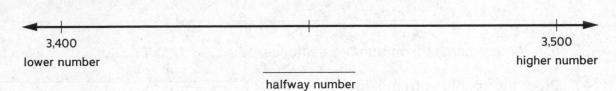

3,400
lower number

_____
halfway number

3,500
higher number

b.

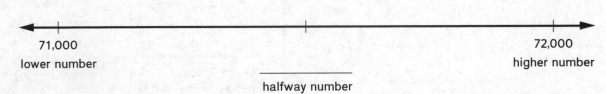

71,000
lower number

_____
halfway number

72,000
higher number

③ Make up a problem of your own.

_____

_____

_____

_____

_____
lower number

_____
halfway number

_____
higher number

Rounding Data

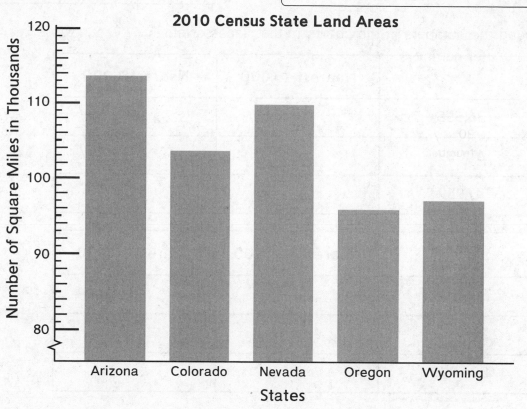

2010 Census State Land Areas

① Use the bar graph to round the data for each state to the nearest hundred-thousand, ten-thousand, and thousand.

	Hundred-Thousand	Ten-Thousand	Thousand
Arizona			
Colorado			
Nevada			
Oregon			
Wyoming			

② Would you choose to round the square miles to hundred-thousands, ten-thousands, or thousands if you were comparing data about these states? Explain.

_____

_____

_____

Rounding Whole Numbers

① Round the numbers in the charts to the places given.

	Nearest 10,000	**Nearest 1,000**
36,555		
61,089		
47,959		

	Nearest 100,000	**Nearest 10,000**
609,909		
362,557		
894,999		

② List in order the steps you might take to round a number to a given place value.

_____

_____

_____

_____

_____

_____

_____

Rounding

① Round the seating capacities in the table below to the nearest thousand.

Women's National Basketball Association (WNBA) Seating Capacity of Home Courts		
Team	**Seating Capacity**	**Rounded to the Nearest 1,000**
Chicago Sky	17,500	
Connecticut Sun	9,518	
Indiana Fever	18,165	
Los Angeles Sparks	13,141	
Minnesota Lynx	19,356	
Phoenix Mercury	18,422	
Seattle Storm	17,072	
Tulsa Shock	17,839	
Washington Mystics	20,308	

Source: www.wnba.com

② Look at your rounded numbers. Which teams' arenas have about the same capacity?

_____

③ Round the population figures in the table below to the nearest hundred-thousand.

U.S. States with the Five Smallest Populations (2010 Census)		
State	**Population**	**Rounded to the Nearest 100,000**
Wyoming	563,626	
Vermont	626,011	
North Dakota	699,628	
Alaska	731,449	
South Dakota	833,354	

Practice

④ _____ = 60 + 60   ⑤ _____ = 54 + 59   ⑥ 185 + 366 = _____

Comparing and Rounding Numbers

Round 1

	Partner 1	<, >, or =	Partner 2
Number			
Rounded to the nearest ten-thousand			
Rounded to the nearest hundred-thousand			

Round 2

	Partner 1	<, >, or =	Partner 2
Number			
Rounded to the nearest ten-thousand			
Rounded to the nearest hundred-thousand			

- -

Round 1

	Partner 1	<, >, or =	Partner 2
Number			
Rounded to the nearest ten-thousand			
Rounded to the nearest hundred-thousand			

Round 2

	Partner 1	<, >, or =	Partner 2
Number			
Rounded to the nearest ten-thousand			
Rounded to the nearest hundred-thousand			

Ordering Cities by Population

Use the population information on *Student Reference Book*, page 282 to complete these problems.

SRB
81

① Find all the cities with a population of less than 500,000. List those cities in order according to population from greatest to least.

City	Population

More To Do

② Find all the cities with a population of more than 1,000,000. List those cities in order according to population from greatest to least.

City	Population

17

Professional Sports Attendance

The table below shows the attendance for various 2013–2014 professional sports teams. Use the table and a place-value tool to answer the questions.

SRB
81,
85-87

	Chicago*	New York*†	Philadelphia	Boston	Washington
Hockey	927,545	738,246	813,411	720,165	740,240
Baseball	2,882,756	3,542,406	3,565,718	3,043,003	2,370,794

Source: ESPN NHL Attendance report 2013–2014 and ESPN MLB Attendance report 2012
*Baseball attendance is for the Chicago Cubs and the New York Yankees.
†Hockey attendance is for the New York Rangers.

(1) Which sport had the greater attendance? _____

(2) Round the attendance at the hockey games.

	Nearest 100,000	Nearest 10,000
Chicago		
New York		
Philadelphia		
Boston		
Washington		

(3) Round the attendance for each baseball team to the nearest million.

Chicago: _____   New York: _____

Philadelphia: _____   Boston: _____

Washington: _____

(4) List the cities in order from greatest to least hockey attendance.

_____

(5) Write a number sentence comparing the greatest and least baseball attendances. Use <, >, or =.

_____

Practice

(6) 210 − 150 = _____   (7) 140 − 80 = _____   (8) 93 − 58 = _____

Finding the Number of Books

The teachers at Forest View School held a reading contest. All students would receive a free book if they read a combined total of 20,000 books by the end of the year.

Quarter	Number of Books Read
First	6,520
Second	5,870
Third	4,460
Fourth	4,780

① Have the students read enough books? _____

② Describe the strategy you used to solve the problem.

_____

_____

✂ -

Finding the Number of Books

Lesson 1-5

NAME   DATE   TIME

The teachers at Forest View School held a reading contest. All students would receive a free book if they read a combined total of 20,000 books by the end of the year.

Quarter	Number of Books Read
First	6,520
Second	5,870
Third	4,460
Fourth	4,780

① Have the students read enough books? _____

② Describe the strategy you used to solve the problem.

_____

Rounding with Base-10 Blocks

You can use base-10 blocks to help you **round** numbers.

Example: Round 64 to the nearest ten.

- Build a model for 64 with base-10 blocks.

- *Think:* What **multiples of 10** are nearest to 64?
  If I take the ones (cubes) away, I would have **60.**
  If I add more ones to make the next ten, I would have **70.**

- Build models for 60 and 70.

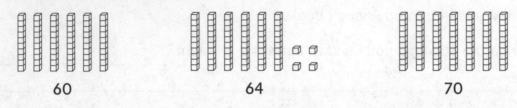

60   64   70

Think: Is 64 closer to 60 or 70? 64 is closer to 60. So, 64 rounded to the nearest ten is 60.

Build models to help you choose the closer number.

1. Round 87 to the nearest ten.

   List the three numbers you will build models for: _____, _____, _____

   87 is closer to _____. So, 87 rounded to the nearest ten is _____.

2. Round 43 to the nearest ten.

   List the three numbers you will build models for: _____, _____, _____

   43 is closer to _____. So, 43 rounded to the nearest ten is _____.

3. Round 138 to the nearest ten.

   List the three numbers you will build models for: _____, _____, _____

   138 is closer to _____. So, 138 rounded to the nearest ten is _____.

4. Round 138 to the nearest *hundred*.

   List the three numbers you will build models for: _____, _____, _____

   138 is closer to _____. So, 138 rounded to the nearest hundred is _____.

Planning a Meal

	Food Choice	Estimated Cost
Protein (meat, beans, tofu)		
Grain (pasta, bread, rice)		
Fruit		
Vegetable		

Total estimated cost: _____

Using Estimation Strategies

Family Note Today students explored different ways of estimating: **rounding** (in which all numbers are rounded to a particular place value), **front-end estimation** (all digits to the right of the greatest place value become zeros), and using **close-but-easier numbers** (numbers are rounded to a number that is close in value and easy to work with). While all methods of estimation are equally valid, some may be more helpful than others for answering specific kinds of questions.

Read the number stories. Choose an appropriate estimation strategy.

SRB
82-89

1. On the walk home from school, Meg stopped at the library for 22 minutes and at her grandmother's house for 38 minutes. She spent 17 minutes walking. She left at 3:00 and was supposed to be home by 4:00.

    a. Did Meg make it home on time? _____ How did you get your answer?

    _____

    _____

    b. Why did you choose your estimation strategy? _____

    _____

    _____

2. You and two friends need to make 100 tacos for a party. You have made 31 tacos. Your friend Chris has made 24 tacos. Your friend Pat thinks he needs to make at least 60 tacos to have enough for the party.

    a. Is Pat correct? _____ How did you get your answer?

    _____

    _____

    b. Why did you choose your estimation strategy? _____

    _____

    _____

Practice

3. $31 + 51 =$ _____
4. $45 + 64 =$ _____
5. $252 + 144 =$ _____

Situation Diagrams for Number Stories

Situation diagrams can help you organize the information in a number story and can help you decide what to do to solve the problem.

SRB 47, 83

Decide which diagram to use for each problem. Complete the diagram. Then solve the problem.

(1) 283 students attended the football game. 371 students attended the soccer game. How many more students attended the soccer game? _____ students

Total	
Part	**Part**

Start **Change** End

Number model:

_____

Quantity
Quantity

Difference

(2) Shawn had $145 in his bank account. He took out some money to buy a new bike. Now he has $85 in his account. How much did his bike cost? _____

Total	
Part	**Part**

Start **Change** End

Number model:

_____

Quantity
Quantity

Difference

(3) Aldo bought milk for 55 cents and a peanut butter and jelly sandwich for 125 cents. How much money did he spend? _____

Total	
Part	**Part**

Start **Change** End

Number model:

_____

Quantity
Quantity

Difference

23

Solving Number Stories

Estimate. Then solve each number story.

SRB
82-84

(1) Paula and her brothers Quinn and Scott want to buy a video game. Quinn has saved $21. Paula has saved $13. Scott has saved $5. The video game they want costs $42 including tax. How much more money do they need to save before they can purchase the video game?

Estimate: About $ _____

Answer: $ _____

Number model with answer: _____

(2) Carter is going away to college and is giving his collection of 531 baseball cards to his cousins. If he gives 227 cards to Lewis, 186 cards to Benny, and 18 cards to Swen, how many are left over?

Estimate: About _____ baseball cards

Answer: _____ baseball cards

Number model with answer: _____

(3) Jeb's family took four days to drive to a national park about 614 miles away. They traveled 173 miles one day, 206 miles the next day, and 51 miles the third day. About how far did they travel on the fourth day to reach the national park?

Estimate: About _____ miles

Answer: _____ miles

Number model with answer: _____

Does your answer make sense? _____ How do you know?

_____

_____

Animal Number Stories

Estimate. Then solve each number story.

SRB
82-84

(1) The zoo needs to move four animals in a truck that can carry only 700 pounds. A leopard can weigh up to 176 pounds. A warthog can weigh up to 250 pounds. A chimpanzee can weigh as much as 130 pounds. What is the maximum weight that the fourth animal can be?

Estimate: About _____ pounds

Answer: _____ pounds

Number model with answer: _____

Does your answer make sense? _____ How do you know?

_____

_____

(2) The combined weight of a mountain lion, an orangutan, and a wolf can be as much as 491 pounds. If the wolf weighs 175 pounds and the orangutan weighs 180 pounds, how much do *two* mountain lions weigh?

Estimate: About _____ pounds

Answer: _____ pounds

Number model with answer: _____

Does your answer make sense? _____ How do you know?

_____

_____

Source: maximum animal weights from www.nationalgeographic.com

Practice

(3) $5 + 8 =$ _____   **(4)** $9 + 6 =$ _____   **(5)** $70 + 50 =$ _____

(6) _____ $= 80 + 50$ **(7)** $67 + 94 =$ _____   **(8)** _____ $= 425 + 275$

25

Number-Tile Addition Problems

Cut out the 0–9 number tiles at the bottom of the page. Use them
to help you solve the problems. Each tile can only be used once.

① Use odd-numbered tiles 1, 3, 5, 7,
and 9 to make the smallest sum.

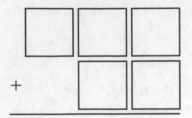

② Use even-numbered tiles 0, 2, 4, 6,
and 8 to make the largest sum.

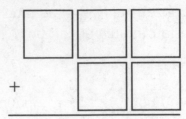

③ Use number tiles 0–9 to complete these four addition problems.
Do not use 0 as the first digit in a number. Use each tile only once.

a.
```
    6  ☐
+ ☐  7
───────
  8  6
```

b.
```
  2  ☐  3
+ ☐  7  1
─────────
  9  5  ☐
```

c.
```
   7  ☐  4
+ ☐  1  5
──────────
1, 2  8  9
```

d.
```
   1  ☐  4
+ ☐  8  ☐
──────────
 ☐  1  4
```

0	1	2	3	4	5	6	7	8	9
0	1	2	3	4	5	6	7	8	9

U.S. Traditional Addition

Family Note In today's lesson students were introduced to U.S. traditional addition. The steps are listed below.

Step 1

Add the 1s: 9 + 7 = 16.

16 ones = 1 ten and 6 ones

Write 6 in the 1s place below the line.

Write 1 above the digits in the 10s place.

```
  1
  7 9
+ 4 7
─────
    6
```

Step 2

Add the 10s: 7 + 4 + 1 = 12.

12 tens = 1 hundred + 2 tens

Write 2 in the 10s place below the line.

Write 1 in the 100s place below the line.

```
  1
  7 9
+ 4 7
─────
1 2 6
```

Make an estimate. Write a number model to show what you did. Then solve using U.S. traditional addition. Compare your answer with your estimate to see if your answer makes sense.

①
```
   3 6
 + 4 6
```

Estimate: _____

②
```
   4 7
 + 9 5
```

Estimate: _____

③ 784 + 889 =

Estimate: _____

④
```
   6 8 9
 + 8 3 9
```

Estimate: _____

⑤ 279 + 1,795 =

Estimate: _____

⑥ 3,746 + 6,255 =

Estimate: _____

Practice

⑦ Round 2,787 to the nearest . . .

hundred _____   thousand _____

⑧ Round 54,681 to the nearest . . .

thousand _____   ten-thousand _____

Packing Bagels

Use centimeter cubes or other small counters to represent bagels. Cut out the boxes below.

For each number of bagels on a tray listed in the table on journal page 19, fill the correct number of boxes with small counters. Fill the largest boxes possible first. Make sure that each box is full. Then fill the next largest size boxes.

When you have figured out how many of each size box you need, write your answers on the journal page.

Boxes of 25 Bagels

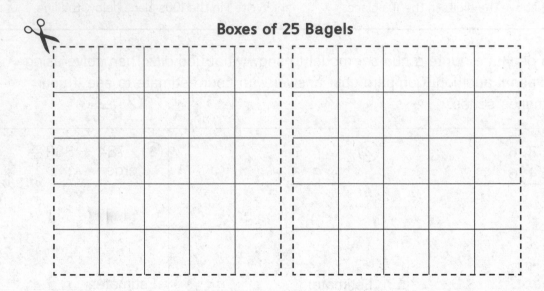

Boxes of 5 Bagels

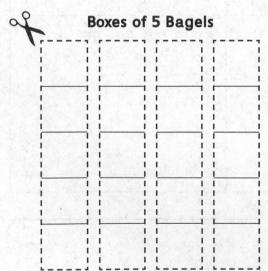

Boxes of 1 Bagel

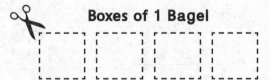

Cracking the Muffin Code

Marcus takes orders at the Marvelous Muffin Market. The muffins are packed into boxes that hold 27, 9, 3, or 1 muffin. Marcus always fills the *largest* box first, uses the *fewest* number of boxes possible, and always sends boxes that are *full*.

When a customer asks for muffins, Marcus fills out an order form.

Hints

For an order of 5 muffins, Marcus writes:

For an order of 19 muffins, Marcus writes:

For an order of 34 muffins, Marcus writes:

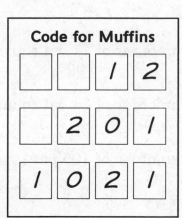

Code for Muffins

		1	2
	2	0	1
1	0	2	1

(1) For an order of 32 muffins, what would Marcus write on the order form?

Explain or show how you know.

(2) How many muffins are in an order with this code?

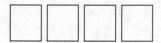

2	0	0	2

Show your work.

Cracking the
Muffin Code (continued)

(3) Oat squares are packed in boxes that hold 100, 10, or 1 square. Marcus uses a similar, but different, coding system for oat squares. Finish coding the orders that came in this morning. The first one is done for you.

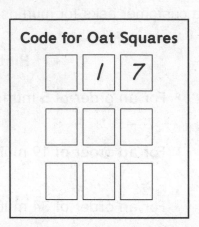

Code for Oat Squares

For an order of 17 oat squares, Marcus writes:

For an order of 24 oat squares, Marcus writes:

For an order of 105 oat squares, Marcus writes:

(4) Explain how using the code for the oat squares is like using place value in our base-10 number system.

(5) Explain how the code for oat squares is like the code for muffins. How are they different?

30

Packing Muffins

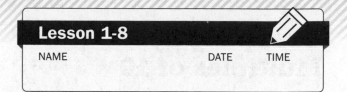

Use centimeter cubes or other small counters to represent muffins. Cut out the boxes of muffins below.

For Problems 1 and 2 on *Math Masters*, page 29, fill the correct number of boxes with counters. Fill the largest boxes possible first. Make sure that each box is full. Then fill the next largest size boxes.

When you have figured out how many of each size box you need for Problem 1 and the number of muffins ordered for Problem 2, write your answers on page 29.

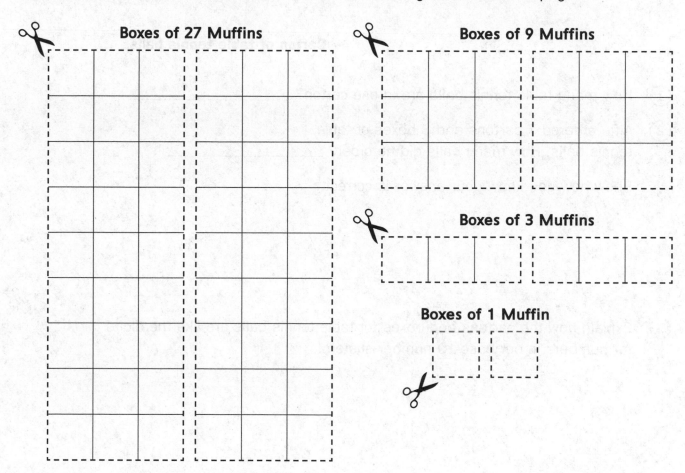

Grouping by Multiples of 10

Alfie is ordering table tennis balls for the recreation center. A box holds 10 balls. A carton of table tennis balls holds 10 boxes.

Box of table tennis balls

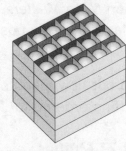

Carton of table tennis balls

(1) How many table tennis balls are in one carton? _____

(2) Alfie ordered 7 cartons and 3 boxes of table tennis balls. How many balls did he order? _____

Show how you know your answer is correct.

(3) Explain how the cartons and boxes for table tennis balls are like the digits for numbers in our base-10 number system.

Practice

(4) 440 + 294 = _____   (5) 166 + 707 = _____

(6) _____ = 425 + 886   (7) 1,474 + 529 = _____

Number-Tile
Subtraction Problems

Cut out the 0–9 number tiles at the bottom of the page. Use them to help you solve the problems. Each of the 20 tiles can be used only one time on this page. Do not use 0 as the first digit in a number.

① Use odd-numbered tiles 1, 3, 5, 7, and 9 to make the smallest difference.

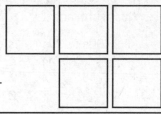

② Use even-numbered tiles 0, 2, 4, 6, and 8 to make the largest difference.

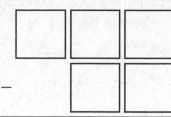

③ Use the other set of number tiles 0–9 to complete these four subtraction problems. Use each tile only once.

a.

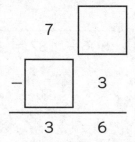

b.

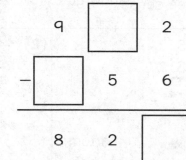

c.

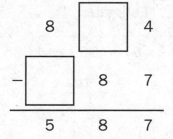

d.

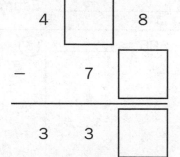

0	1	2	3	4	5	6	7	8	9
0	1	2	3	4	5	6	7	8	9

U.S. Traditional Subtraction

SRB
82-84,
100-101

Family Note In today's lesson students were introduced to U.S. traditional subtraction. The process is shown below for the problem 653 – 387.

Step 1:

Start with the ones. Trade 1 ten for 10 ones. Subtract the ones.

100s	10s	1s
	4	13
6	5̶	3̶
− 3	8	7
		6

Step 2:

Go to the tens. Trade 1 hundred for 10 tens. Subtract the tens.

100s	10s	1s
	14	
5	4̶	13
6̶	5̶	3̶
− 3	8	7
	6	6

Step 3:

Go to the hundreds. We don't need to regroup, so just subtract.

100s	10s	1s
	14	
5	4̶	13
6̶	5̶	3̶
− 3	8	7
2	6	6

Make an estimate. Write a number model to show what you did. Then solve using U.S. traditional subtraction. Compare your answer with your estimate to see whether your answer makes sense.

①
```
   8 5
 − 3 8
```

Estimate: _____

②
```
   6 1 3
 − 2 4 9
```

Estimate: _____

③ 506 − 187 = _____

Estimate: _____

④ 951 − 695 = _____

Estimate: _____

⑤
```
   1, 5 4 4
 −    7 4 9
```

Estimate: _____

⑥ 7,003 − 4,885 =

_____

Estimate: _____

Practice

⑦ 740 + 294 = _____

⑧ 2,566 + 807 = _____

Personal References for Inches, Feet, and Yards

1. Record your personal references below.

Units of Measure	Personal References
1 inch (in.)	
1 foot (ft)	
1 yard (yd)	

2. Record your estimates and measurements of objects.

Object or Distance	My Estimate	My Measurement
	About _____ in.	About _____ in.
	About _____ in.	About _____ in.
	About _____ ft	About _____ ft
	About _____ ft	About _____ ft
	About _____ yd	About _____ yd
	About _____ yd	About _____ yd

Migratory Bird Flights

According to the Smithsonian National Zoological Park, migrating birds usually choose the altitude at which they fly based on where the best wind conditions are found. Sometimes this involves setting records for flight altitudes. Bar-headed geese are known to cross the Himalayas at 29,500 feet. A Ruppell's griffon vulture has set the world record at 37,000 feet. A mallard holds the record for the highest documented flight altitude for a bird in North America at 21,000 feet.

① Altitudes are usually recorded in feet. What would the table below show if we also recorded the heights in inches, yards, and miles? Use a calculator to help fill in the table. Record your data to the nearest whole number.

Migratory Bird Flight Heights				
Bird	**In Inches**	**In Feet**	**In Yards**	**In Miles**
Bar-headed goose		29,500		
Ruppell's griffon vulture		37,000		
Mallard		21,000		

② Write equations to show how you converted the mallard's altitude from feet to inches, yards, and miles.

To inches: _____

To yards: _____

To miles: _____

③ Look at the data in the table. Do you agree that altitude for migrating birds is best reported in feet? Why or why not?

_____

_____

Snake Lengths

Use the measurement scales to solve the problems.

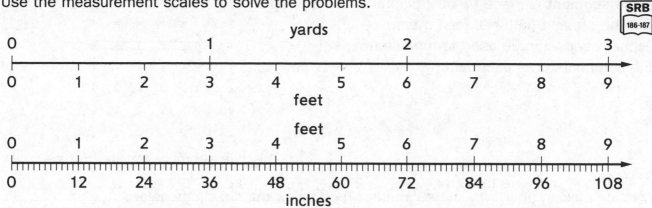

SRB
186-187

①

Feet	Inches
1	
6	
8	
12	

②

Yards	Feet
1	
3	
8	
16	

③ The king cobra can measure a little over 4 yards in length. The black mamba can reach a length of almost 5 yards. What is the combined length of the two snakes in feet?

Answer: _____ feet

④ The Burmese python can be anywhere from 16 to 23 feet long. What is the difference in length in inches between the longest and shortest Burmese python?

Answer: _____ inches

Practice

⑤ Write 4,857 in words.

_____

⑥ Write 14,066 in words.

_____

37

Modeling Line Segments

A **line segment** is made up of 2 points and the straight path between them. Rubber bands can be used to represent line segments on a geoboard.

Example:

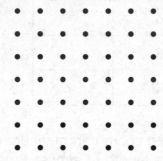

This line segment touches 5 pins.

Practice making your own line segments. Then follow the directions below. Record your work on the geoboard diagrams by drawing the position of the rubber band.

① Make a line segment that touches 4 pins.

② Make a line segment that touches 4 different pins.

③ Make the shortest line segment possible.

④ Make the longest line segment possible.

⑤ Cami says she cannot make a **line** on her geoboard. Do you agree? Explain why or why not. (*Hint:* Look up **line** in the glossary of your *Student Reference Book*.)

_____

_____

Solving a Collinear-Points Puzzle

Three or more points on the same line are called **collinear points.**

Example: Points *A*, *B*, and *C* are collinear points on the line below. *ABC* means that *A*, *B*, and *C* are collinear points, and point *B* is between points *A* and *C*.

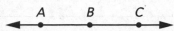

① The following are true statements about $\overleftrightarrow{EH}$:
EFH and *FGH*

The following are false statements about $\overleftrightarrow{EH}$:
FEH and *FHG*

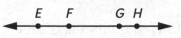

   a. Name two more true statements about line *EH*. _____

_____

   b. Name two more false statements about line *EH*. _____

_____

② Place collinear points *J, K, L, M, N,* and *O* on the line below using these clues:

• *J* and *O* are not between any points.

• *MKL*

• *NLJ*

• *MKN*

③ Show a different solution to the puzzle in Problem 2.

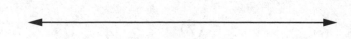

④ Create a collinear-points puzzle on the back of this page. Be sure to give enough clues. Record your solution on the line below. Ask someone to solve your puzzle. Can the problem solver find more than one solution to your puzzle?

Line Segments, Lines, and Rays

① List at least 5 things in your home that remind you of line segments.

SRB
226-227,
230-231

_____

_____

_____

Use a straightedge to complete Problems 2 and 3.

② **a.** Draw and label line *EF*.   **b.** Draw and label line segment *EF*.

c. Explain how your drawings of line *EF* and line segment *EF* are different.

_____

_____

③ **a.** Draw and label ray *SR*.

b. Anita says ray *SR* can also be called ray *RS*. Do you agree? Explain.

_____

_____

④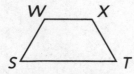

Name the parallel line segments.

_____

Practice

⑤
```
    9 6 4
  - 3 4 8
```

⑥
```
    6 6 2
  - 4 9 7
```

⑦
```
   2, 4 2 3
 - 1, 4 9 1
```

Pattern-Block Sort

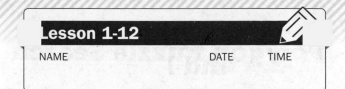

Label one sheet of paper: **These fit the rule.**

Label another sheet of paper: **These do NOT fit the rule.**

Sort the pattern blocks (hexagon, trapezoid, square, triangle, 2 rhombuses) according to the rules given below. Then use the shapes marked "PB" on your Geometry Template to record the results of your sort.

① Exactly 4 sides		② All sides the same length	
These fit the rule.	These do NOT fit the rule.	These fit the rule.	These do NOT fit the rule.

③ 4 sides *and* all sides the same length		④ All angles the same measure, or size, *and* all sides the same length	
These fit the rule.	These do NOT fit the rule.	These fit the rule.	These do NOT fit the rule.

⑤ Make up your own rule. Sort the pattern blocks according to your rule. Record your rule and the pattern blocks that fit your rule on the back of this page.

Polygon Puzzle Search

(1) Study the figure at the right.

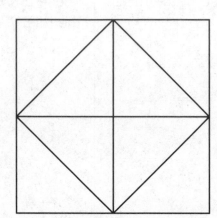

SRB
232-234

    a. How many triangles do you see?

    _____

    b. How many triangles have
    a right angle?

    _____

(2) Study the figure at the right.

    a. How many squares do you see?

    _____

    b. How many right triangles?

    _____

    c. How many rectangles
    that are *not* squares?

    _____

(3) Make up a geometry puzzle like the one in Problem 2. Use a straightedge
to draw line segments to connect some of the dots in the array.
Write the answers on the back of this page. Then ask someone to count
the number of different polygons in your puzzle.

    •   •   •   •

    •   •   •   •

    •   •   •   •

    •   •   •   •

Angles and Quadrilaterals

Use a straightedge to draw the geometric figures.

SRB
228-229,
233

(1) Draw 2 examples of a rectangle.

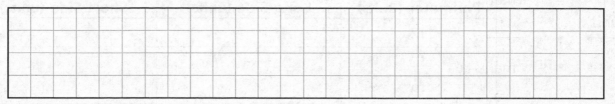

(2) Draw 2 examples of a right triangle.

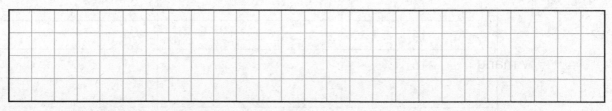

(3) How are the shapes in Problems 1 and 2 similar? How are they different?

_____

_____

(4) **a.** Draw right angle *DEF*.

(5) Draw an angle that is larger than a right angle. Label the vertex *K*.

b. What is the vertex of the angle? Point _____

c. What is another name for ∠*DEF*? _____

Practice

Use U.S. traditional subtraction.

(6) _____ = 756 − 348

(7) 700 − 450 = _____

(8) 7,942 − 3,887 = _____

Geoboard Perimeters

On a geoboard, make rectangles or squares with the perimeters given below.
Record the lengths of the long side and short side of each shape.

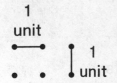

Perimeter (units)	Long Side (units)	Short Side (units)
12		
12		
12		
14		
14		
14		
16		
16		
16		
16		

Geoboard Perimeters

On a geoboard, make rectangles or squares with the perimeters given below.
Record the lengths of the long side and short side of each shape.

Perimeter (units)	Long Side (units)	Short Side (units)
12		
12		
12		
14		
14		
14		
16		
16		
16		
16		

Pattern-Block Perimeters

① Use the following pattern blocks to create shapes with as many *different* perimeters as you can: 1 hexagon, 3 trapezoids, 3 blue rhombuses, and 3 triangles.

- Every shape must include all 10 pattern blocks.

- Each side of a pattern block measures 1 unit. The long side of a trapezoid pattern block measures 2 units.

- At least one side of every pattern block must *line up exactly* with a side of another pattern block. See figures.

 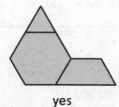 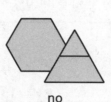 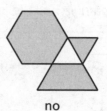 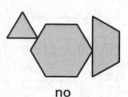

    yes   yes   no   no   no

② Use your Geometry Template to record your shapes on a separate sheet of paper. The polygons should all have different perimeters. Write the perimeter next to each shape.

③ What was the smallest perimeter you were able to make? _____ units
Describe the strategy you used to find this perimeter.

_____

_____

_____

④ What was the largest perimeter you were able to make? _____ units
Describe the strategy you used to find this perimeter.

_____

_____

_____

Finding the Perimeter

Family Note In class, students developed some rules, or *formulas,* for finding the perimeter of a rectangle. Here are three possible formulas:

- Add the measures of the four sides: perimeter of a rectangle = length + length + width + width. This formula can be abbreviated as: $p = l + l + w + w$.

- Add the two given sides and double the sum: perimeter of a rectangle = 2 * (length + width). This formula can be abbreviated as: $p = 2 * (l + w)$.

- Double the length, double the width, and then add: perimeter of a rectangle = (2 * length) + (2 * width). This formula can be abbreviated as: $p = 2l + 2w$.

In all of the formulas, the letter *p* stands for the *perimeter of a rectangle,* the letter *l* stands for the *length of the rectangle,* and the letter *w* stands for the *width of the rectangle.*

Find the perimeters of the rectangles below.

SRB
200

(1)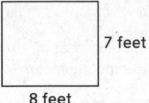

7 feet

8 feet

Perimeter: _____ feet

(2)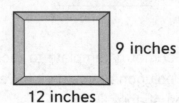

9 inches

12 inches

Perimeter: _____ inches

(3)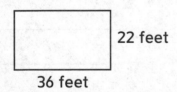

22 feet

36 feet

Perimeter: _____ feet

(4)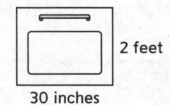

2 feet

30 inches

Perimeter: _____ inches

(5) The perimeter of a garden is 42 feet. The length is 15 feet. What is the width?

Width: _____ feet

Practice

Round each number to the nearest ten-thousand and hundred-thousand.

(6) 421,492 _____ _____

(7) 895,531 _____ _____

46

Unit 2: Family Letter

Multiplication and Multiplicative Comparison

In Unit 2 students build on prior work multiplying whole numbers. The focus is on multiplication in a variety of contexts, including rectangular-array patterns and work with factors, factor pairs, multiples, prime numbers, and composite numbers.

This unit introduces the concept of multiplicative comparison, or using multiplication to compare one quantity to another. Take the following number story: *Mike earned $4. Sue earned 7 times as much as Mike.* Here Sue's earnings are compared to Mike's as being 7 times as much. Based on this comparison, we can find how much Sue earned ($4 * 7 = $28).

Measurement work in Unit 2 is tied to multiplication. Working with units of time, students multiply to convert from hours to minutes and minutes to seconds. They are introduced to the area formula for rectangles, $A = l * w$, in which A is area, l is length, and w is width.

Applying the formula for the area of a rectangle:

$A = 15\ \text{cm}^2$ $w = 3\ \text{cm}$

$l = 5\ \text{cm}$

Students also work with patterns found in square numbers, multiples, factors, and "What's My Rule?" tables. They practice looking more deeply into patterns by identifying ones that are apparent but are not stated in the rule. For example, students may notice in the pattern based on the rule *multiply a number by itself* that every other square number is even.

Classifying Geometric Figures; Symmetry

Students build on their study of geometry in Unit 1 by identifying properties of shapes. They explore the properties of angles and triangles by identifying right, obtuse, and acute angles in triangles. Students begin work with classification, an important geometry skill, by sorting quadrilaterals according to the number of pairs of parallel sides.

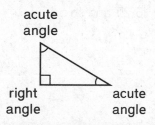

acute angle

right angle   acute angle

Identifying properties of right triangles

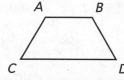

The trapezoid has one pair of parallel sides: $\overline{AB}$ and $\overline{CD}$.

Symmetry is another focus in Unit 2. Symmetry is found in natural objects like flowers, insects, and the human body, as well as in buildings, furniture, clothing, and paintings.

Please keep this Family Letter for reference as your child works through Unit 2.

Vocabulary

Important terms in Unit 2:

array An arrangement of objects in a regular pattern, usually in rows and columns.

composite number A counting number that has more than two different *factors*. For example, 4 is a composite number because it has three *factors*: 1, 2, and 4.

factor One of two or more numbers that are multiplied to give a *product*. For example, 4 * 5 = 20; so 20 is the *product*, and 4 and 5 are the factors.

formula A general rule for finding the value of something. A formula is often written using letters to stand for the quantities involved. For example, the formula for the area of a rectangle may be written as $A = l * w$, where A represents the area of the rectangle, l represents its length, and w represents its width.

line of symmetry A line drawn through a figure that divides the figure into two parts that are mirror images of each other. The two parts look alike but face in opposite directions.

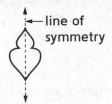

line of symmetry

line symmetry A figure has line symmetry if a line can be drawn dividing it into two parts that are mirror images of each other. The two parts look alike but face in opposite directions.

multiple A *product* of a number and a counting number. The multiples of 7, for example, are 7, 14, 21, 28, and so on.

prime number A counting number greater than 1 that has exactly two *factors*: itself and 1. For example, 5 is a prime number because its only *factors* are 5 and 1.

product The result of multiplying two numbers called *factors*. For example, in 4 * 3 = 12, the product is 12.

rectangular array An arrangement of objects into rows and columns that form a rectangle. All rows and columns must be filled. Each row has the same number of objects and each column has the same number of objects.

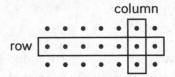

square array An arrangement of objects into rows and columns that form a square. All rows and columns must be filled. All of the rows and all of the columns have the same number of objects, making the number of rows and columns equal. A square array can be a representation of a *square number*.

square number A number that is the product of a counting number multiplied by itself. For example, 25 is a square number because 5 * 5 = 25. The square numbers are 1, 4, 9, 16, 25, and so on.

Do-Anytime Activities

To work with your child on concepts taught in this unit, try these activities:

1. Ask your child to list the first 5 or 10 multiples of different 1-digit numbers.

2. Help your child recognize real-world examples of right angles, such as the corner of a book, and of parallel lines, such as railroad tracks.

3. Encourage your child to look for symmetrical objects and if possible to collect pictures of symmetrical objects from magazines and newspapers. For example, the right half of the printed letter T is the mirror image of the left half.

Building Skills through Games

In this unit your child will play the following games to develop understanding of factors and multiples.

Buzz* and *Bizz-Buzz See *Student Reference Book,* page 252. *Buzz* provides practice finding multiples of whole numbers. *Bizz-Buzz* provides practice finding common multiples of two whole numbers.

As You Help Your Child with Homework

As your child brings assignments home, you may want to go over the instructions together, clarifying them as necessary. The answers listed below will guide you through the Home Links for this unit.

Home Link 2-1

1. $2 * 2$; 16; $5 * 5$; $6 * 6$

3. Sample answers: The product of two even factors is even; the product of two odd factors is odd.

5 **a.** $5 * 5 = 25$ **b.** $5 * 5 = 25$ shows the same number of rows and columns.

7. 11; 37; 63; $+ 26$

Home Link 2-2

1.

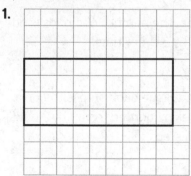

$9 * 4 = 36$; 36

3. $8 * 6 = 48$; 48   5. $9 * 6 = 54$; 54

7. 47

Home Link 2-3

1. **9:** $1 * 9 = 9, 9 * 1 = 9, 3 * 3 = 9$; 1 and 9, 3 and 3;
10: $1 * 10 = 10, 10 * 1 = 10, 2 * 5 = 10, 5 * 2 = 10$;
1 and 10, 2 and 5; **17:** $1 * 17 = 17, 17 * 1 = 17$; 1 and
17; **40:** $1 * 40 = 40, 2 * 20 = 40, 4 * 10 = 40$,
$5 * 8 = 40, 8 * 5 = 40, 10 * 4 = 40, 20 * 2 = 40$,
$40 * 1 = 40$; 1 and 40, 2 and 20, 4 and 10, 5 and 8

3. 2,863   5. 2,182

Home Link 2-4

1. 4, 8, 12, 16, 20

3. **a.** 3, 6, 9, 12, 15, 18, 21, 24, 27, 30

   b. 5, 10, 15, 20, 25, 30, 35, 40, 45, 50

   c. 15 and 30

5. No. 35 cannot be divided evenly by 6.

7. 36; 60; 84; $+ 12$ 9. 69; 35; 1; $- 17$

Home Link 2-5

1. 1, ⑪; prime

3. 1, ②, ③, 4, 6, 8, 12, 24; composite

5. 1, ②, ③, 4, 6, 9, 12, 18, 36; composite

7. 1, ②, ⑤, 10, 25, 50; composite

Home Link 2-6

1. 9 grapefruits
3. Answers vary.
5. thirty thousand, forty-one
7. nine million, ninety thousand, five hundred six

Home Link 2-7

1. 240, 480, 660
3. 1,020
4. 47
5. 14,220
7. 7,424
9. 7,298

Home Link 2-8

1. $n = 7 * 9$; 63
3. $32 = 4 * x$; 8
5. 399
7. 2,149

Home Link 2-9

1. Answers vary; Sample answer: $6 * 9 = m$; 54
5. $50,000 + 6,000 + 30 + 7$
7. $700,000 + 10,000 + 6,000 + 300 + 5$

Home Link 2-10

1. C, D
3. C, D
5. C, D, E, F
7. A, B, E, F
9. 1, 2, 3, 4, 6, 12

Home Link 2-11

1. Sample answer:

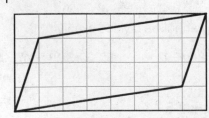

3. Sample answer:

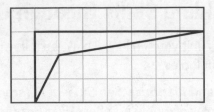

5. 150
7. 480

Home Link 2-12

1.

3. Answers vary.
5. 7,171
7. 2,595

Home Link 2-13

1. 3; 5; 36; 54

Sample answer: If you add the digits of each of the multiples of 9, the sum is 9.

3. a.

| 1 | 2 | 3 | 4 | 5 |

Sample answer: The number of circles is odd and increases by 2 every time.

b. 11; 19

c. Sample answer: Since each step is the next odd number, I skip counted from 1 by 2 until I got to the 10th step.

5. 250,004

Building Arrays

A **rectangular array** is an arrangement of objects in rows and columns. Each row has the same number of objects, and each column has the same number of objects.

Work with a partner to build arrays. For each array, take turns rolling dice. The number of dots on the first die represents the number of rows. Write this number in the table under *Rows*. The number of dots on the second die represents the number of cubes in each row. Write this number under *Columns*. Then use centimeter cubes to build the array on the dot grid. How many cubes are in the array? Write this number under *Array Total* on the table.

Rows	Columns	Array Total

Rows	Columns	Array Total

One More, One Less

① Square number: _____

Equation: _____

New equation: _____

② Square number: _____

Equation: _____

New equation: _____

③ Square number: _____

Equation: _____

New equation: _____

④ Square number: _____

Equation: _____

New equation: _____

⑤ Square number: _____

Equation: _____

New equation: _____

⑥ Describe any patterns you notice.

_____

_____

_____

_____

Exploring Square Numbers

A **square number** is a number that can be written as the product of a number multiplied by itself. For example, the square number 9 can be written as 3 * 3.

$3 * 3 = 9$

① Fill in the missing factors and square numbers.

Factors	Square Number
	4
3 * 3	9
4 * 4	
	25
	36

② What pattern(s) do you see in the factors? In the products?

_____

_____

③ What other pattern(s) do you see in the table?

_____

_____

④ Write an equation to describe each array.

a.

b.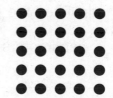

Equation: _____   Equation: _____

⑤ a. Which of the arrays above shows a square number? _____

    b. Explain. _____

Practice

⑥ 32, 45, 58, _____, _____, _____   Rule: _____

⑦ _____, _____, _____, 89, 115, 141   Rule: _____

Finding the Area of a Rectangle

(1) Find the perimeter. You may use a centimeter cube or ruler. Write an equation to show how you calculated the perimeter.

Perimeter: _____ centimeters

Equation: _____

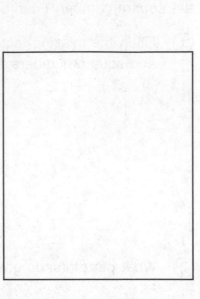

(2) Find the area.

Area: _____ square centimeters

Finding the Area of a Rectangle

(1) Find the perimeter. You may use a centimeter cube or ruler. Write an equation to show how you calculated the perimeter.

Perimeter: _____ centimeters

Equation: _____

(2) Find the area.

Area: _____ square centimeters

Area of a Rectangle

① Take turns rolling a die. The first roll represents the length of a rectangle. The second roll represents the width of the rectangle.

SRB
58-60,
202-204

② Use square pattern blocks to build the rectangle. Count squares to find the area.

Example:

First roll 4, second roll 3

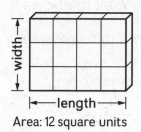

Area: 12 square units

③ Record your results in the table.

First Roll (length)	Second Roll (width)	Area (square units)
4	3	12

④ Describe a pattern in your table.

_____

_____

⑤ Without building the rectangle, can you use this pattern to find the area of a rectangle with a length of 8 units and a width of 7 units? Explain your answer.

_____

_____

Comparing Perimeter and Area

- Cut 6 rectangles that are 6 columns by 7 rows from the centimeter grid paper.

- Record the area and the perimeter of one of these rectangles in Problem 1.

- Divide each rectangle into 3 parts, each with the same number of boxes. Use a different colored pencil to trace each part along the grid lines. The 3 parts do not have to be rectangles, but all the boxes in the rectangles must be used.

- Divide each rectangle in a different way.

① For a rectangle that is 6 cm by 7 cm:

Area = _____ sq. cm   Perimeter = _____

② Record the perimeters for the divisions of the 6 rectangles in the table.

Rectangle	Perimeters		
	Part 1	Part 2	Part 3
1			
2			
3			
4			
5			
6			

③ What is the area for each of the parts? _____

④ **a.** Describe one relationship between perimeter and area.

_____

_____

b. Is the relationship the same for rectangles and irregular figures? Explain.

_____

_____

Area of a Rectangle

① Draw a rectangle that has length of 9 units and width of 4 units.

② Draw a rectangle that has a length of 7 units and a width of 8 units.

SRB
202-204

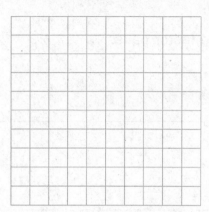

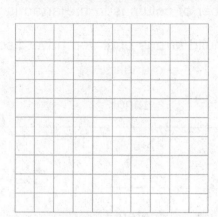

Equation: _____

Area = _____ square units

Equation: _____

Area = _____ square units

Use the formula $A = l * w$ to find the area of each rectangle.

③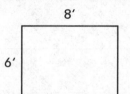
8'
6'

Equation: _____

Area = _____ square feet

④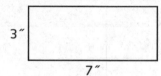
3″
7″

Equation: _____

Area = _____ square inches

⑤ Riley's dining room tabletop is 9 feet long and 6 feet wide. What is the area of the tabletop?

Equation: _____

Area = _____ square feet

Practice

⑥ 368 − 59 = _____

⑦ 194 − 147 = _____

⑧ _____ = 1,729 − 623

Factoring Numbers with Cube Arrays

Use centimeter cubes to build arrays for the numbers shown below. For each array, write the **factor pair.** Remember that the number of rows in the array is one **factor** and the number of columns in the array is the other **factor.**

Continue to build arrays until you have discovered all of the factor pairs.

① 12

Factor pairs: _____

② 6

Factor pairs: _____

③ 15

Factor pairs: _____

④ 24

Factor pairs: _____

⑤ 30

Factor pairs: _____

⑥ 33

Factor pairs: _____

⑦ How can you make sure that you include all of the factors for a number?

_____

_____

_____

_____

Working with Factor Pairs

(1) Write equations to help you find all the factor pairs of each number below. Use dot arrays, if needed.

SRB 53

Number	Equations with Two Factors	Factor Pairs
6	$1 * 6 = 6$ $2 * 3 = 6$ $3 * 2 = 6$ $6 * 1 = 6$	1 and 6 2 and 3
9		
10		
17		
40		

Practice

(2) 356 + 433 = _____

(3) _____ = 2,167 + 696

(4) _____ = 4,578 − 2,232

(5) 3,271 − 1,089 = _____

59

Solving Number Stories

Solve the number stories. Show your work in the space provided. Use counters or centimeter cubes as needed.

1. Mrs. Amador's students are going to work in groups on a project. There can be any number of groups, but each one must have the same number of students and no one may be left out. There are 24 students in Mrs. Amador's class. What are all of the different ways her students could organize themselves into groups?

   _____

   _____

   How do you know that you have found all of the possible groupings?

   _____

   _____

   _____

   _____

   _____

2. Two fourth-grade classes presented their science projects. In Ms. Bauer's class a student presented every 5 minutes. In Mr. Harris's class a student presented every 8 minutes. Both classes started their presentations at the same time. How many minutes passed after the first presentations before both classes had presenters starting at the same time?

   _____ minutes

   Explain how you solved the problem.

   _____

   _____

Finding Multiples

① List the first 5 multiples of 4. _____

② List the first 10 multiples of 2. _____

③ a. List the first 10 multiples of 3. _____

   b. List the first 10 multiples of 5. _____

   c. List the multiples of 3 that are also multiples of 5. _____

④ Is 28 a multiple of 7? _____ Explain. _____

_____

⑤ Is 35 a multiple of 6? _____ Explain. _____

_____

⑥ a. List the factors of 15. List the multiples through 15 of each factor.

Factors of 15	Multiples of the Factors (of 15)

   b. Is 15 a multiple of each of its factors? _____ Explain. _____

_____

Practice

⑦ 24, _____, 48, _____, 72, _____   Rule: _____

⑧ _____, 108, 162, _____, 270, _____   Rule: _____

⑨ 86, _____, 52, _____, 18, _____   Rule: _____

⑩ 425, _____, 339, _____, 253, _____   Rule: _____

Exploring Goldbach's Conjecture

① Write each of the following numbers as the sum of two prime numbers.

Examples: $56 = \underline{43 + 13}$ $26 = \underline{13 + 13}$

a. $6 = \underline{\hspace{3cm}}$ **b.** $12 = \underline{\hspace{3cm}}$

c. $18 = \underline{\hspace{3cm}}$ **d.** $22 = \underline{\hspace{3cm}}$

e. $24 = \underline{\hspace{3cm}}$ **f.** $34 = \underline{\hspace{3cm}}$

The answers to these problems are examples of **Goldbach's Conjecture,** which states that any even number greater than 2 is the sum of two prime numbers. A **conjecture** is something you believe is true even though you cannot be certain that it is because it hasn't been proven. Goldbach's Conjecture appears to be true because no counterexample has ever been found, but no one has ever proven it. Anyone who can either prove or disprove Goldbach's Conjecture will become famous.

② Work with a partner. Write the numbers on the grid on *Math Masters*, page 63 as the sum of two prime numbers.

③ Can any of the numbers on the grid be written as the sum of two prime numbers in more than one way? If so, give an example. Show all possible ways.

_____

_____

_____

Try This

④ Write 70 as the sum of two prime numbers in as many ways as you can.

_____

_____

Exploring Goldbach's Conjecture (continued)

Write each number below as the sum of two prime numbers.

4	8	10	12	13
_2 + 2_	_____	_____	_____	_____
14	16	18	20	22
_____	_____	_____	_____	_____
24	26	28	30	32
_____	_____	_____	_____	_____
34	36	38	40	42
_____	_____	_____	_____	_____
44	46	48	50	52
_____	_____	_____	_____	_____
54	56	58	60	62
_____	_____	_____	_____	_____
64	66	68	70	72
_____	_____	_____	_____	_____
74	76	78	80	82
_____	_____	_____	_____	_____
84	86	88	90	92
_____	_____	_____	_____	_____
94	96	98	100	102
_____	_____	_____	_____	_____

Prime and Composite Numbers

A **prime number** is a whole number that has exactly two different factors—1 and the number itself. A **composite number** is a whole number that has more than two different factors.

For each number:

- List all of its factors.

- Write whether the number is prime or composite.

- Circle all of the factors that are prime numbers.

	Number	Factors	Prime or Composite?
①	11		
②	19		
③	24		
④	29		
⑤	36		
⑥	49		
⑦	50		
⑧	70		
⑨	100		

Practice

Solve.

⑩ $841 + 527 =$ _____

⑪ _____ $= 3,263 + 5,059$

⑫ $7,461 + 2,398 =$ _____

⑬ _____ $= 4,172 - 3,236$

⑭ $8,158 = 5,071 +$ _____

⑮ $3,742 - 3,349 =$ _____

64

Little and Big

Ana has two dogs, Little and Big. She had a picture of them sitting side by side, but the dogs ate most of the picture. Most of Big is now missing!

She used dog treats to measure Little's height in the picture, starting at the floor by his paw and going up to the top of his head. Little's height was 2 dog treats.

Ana measured the picture of Big in the same way before the picture was torn, using the same dog treats. Big's height was 6 dog treats.

① Look at the picture of Little. Measure Little's height using paper clips.

_____ paper clips

② Predict the height of Big in the picture if you measured him using the same paper clips.

_____ paper clips

③ Show or tell how you figured out what to predict. You may use diagrams, words, or calculations. Explain your steps carefully.

Picture of Little

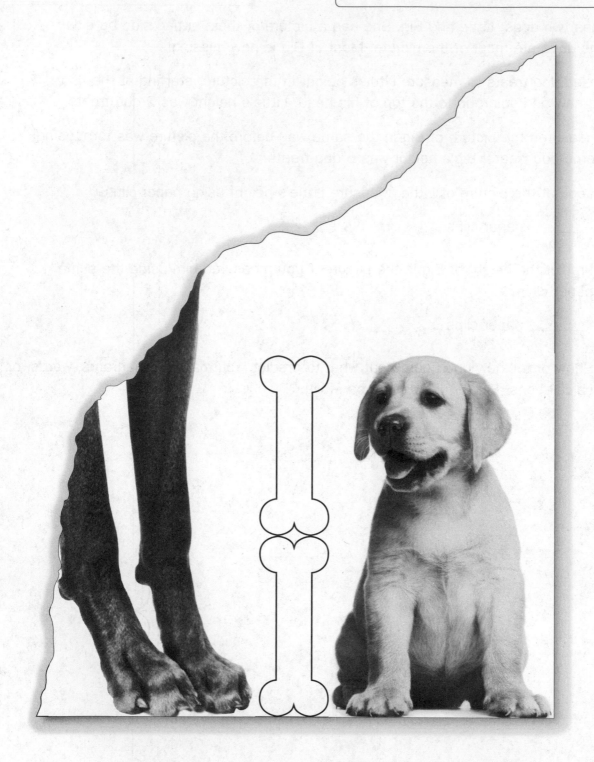

Dog Treat Cutouts

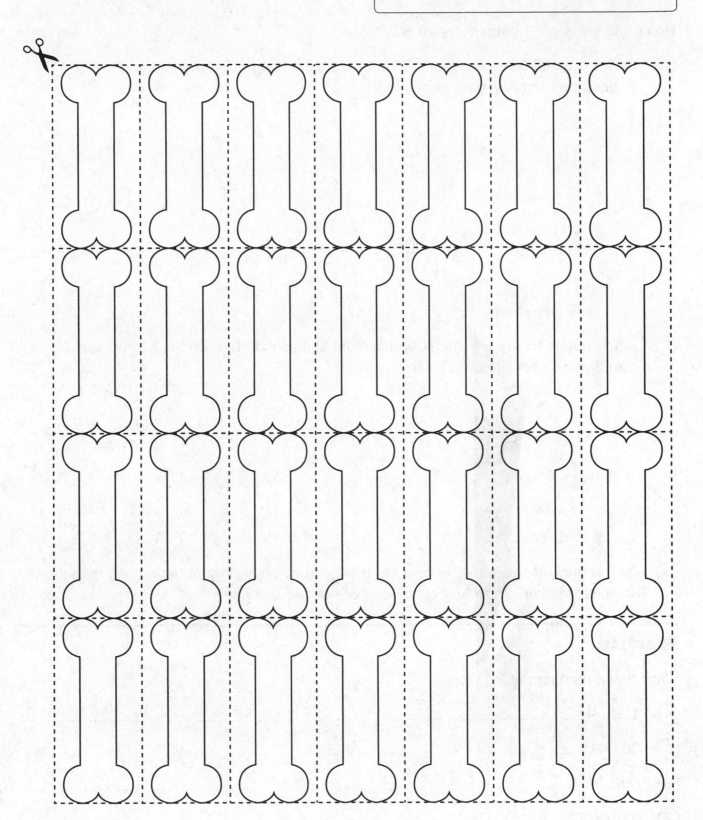

Using Multiplication

Home Market sells 3 grapefruits for $2.

SRB
42, 53

① Darius spent $6 on grapefruits. How many did he buy? Use words, numbers, or diagrams to show your reasoning.

_____ grapefruits

② Jana bought 15 grapefruits. How much did she spend? Use words, numbers, or diagrams to show your reasoning.

_____ dollars

③ On the back of this page, write a multiplication number story about buying grapefruits at Home Market. Show how to solve your number story.

Practice

Write these numbers using words.

④ 12,309 _____

⑤ 30,041 _____

⑥ 600,780 _____

⑦ 9,090,506 _____

How Long Is a Second? a Minute? an Hour?

Fill in the chart below to show what you think you could do in a second, a minute, and an hour.

Second	Minute	Hour

✂ -

How Long Is a Second? a Minute? an Hour?

Fill in the chart below to show what you think you could do in a second, a minute, and an hour.

Second	Minute	Hour

Converting Units of Time

Use the measurement scales to fill in the tables and answer the questions.

minutes

```
0     1     2     3     4     5     6     7     8     9     10
|-----|-----|-----|-----|-----|-----|-----|-----|-----|-----|----->
0    60   120   180   240   300   360   420   480   540   600
seconds
```

hours

```
0     1     2     3     4     5     6     7     8     9     10
|--|--|--|--|--|--|--|--|--|--|--|--|--|--|--|--|--|--|--|--|----->
0  30  60  90 120 150 180 210 240 270 300 330 360 390 420 450 480 510 540 570 600
minutes
```

**① **

Hours	Minutes
1	60
4	
8	
11	

**② **

Minutes	Seconds
1	60
5	
10	
20	

③ Zac worked on his spelling for 9 minutes last night and 8 minutes this afternoon. How many seconds did he work? Answer: _____ seconds

④ Eton's baby sister took a nap for 2 hours and 22 minutes yesterday and 1 hour and 35 minutes today. How many more minutes did she sleep yesterday than today? Answer: _____ minutes

Try This

⑤ How many seconds did Eton's baby sister sleep all together?
Answer: _____ seconds

Practice

⑥ 945 + 1,055 = _____

⑦ 2,953 + 4,471 = _____

⑧ 4,552 + 4,548 = _____

⑨ 3,649 + 3,649 = _____

70

Comparing Statements

Math Message

Look at the ribbons and then read the statement given.

Eve's
ribbon

Maxine's
ribbon

Statement: Eve's ribbon is shorter than Maxine's ribbon.

Measure the two ribbons in centimeters. Write two more statements describing
how the lengths compare.

(1) Statement: _____

(2) Statement: _____

Comparing Statements

Math Message

Look at the ribbons and then read the statement given.

Eve's
ribbon

Maxine's
ribbon

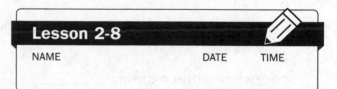

Statement: Eve's ribbon is shorter than Maxine's ribbon.

Measure the two ribbons in centimeters. Write two more statements
describing how the lengths compare.

(1) Statement: _____

(2) Statement: _____

Comparing Animal Weights

The table below shows about how much an animal weighs at birth and about how much its mother weighs.

Animal	Weight of Baby in Kilograms	Weight of Mother in Kilograms
Beluga whale	80	1,000
Blue whale		190,000
Deer	3	
Elephant	115	4,000
Giraffe	75	1,200
Polar bear		450
White rhinoceros	50	2,200

Write an equation with an unknown. Then find the answer and write it in the table above.

① A mother deer weighs 30 times as much as her baby. How much does the mother weigh?

Equation with unknown: _____ Answer: _____ kilogram(s)

② A mother polar bear weighs 450 times as much as her baby. How much does the baby weigh?

Equation with unknown: _____ Answer: _____ kilogram(s)

③ A mother white rhinoceros weighs about how many times as much as her baby?

Equation with unknown: _____ Answer: _____ times
as much

④ A mother blue whale weighs 95 times as much as her baby. How much does her baby weigh?

Equation with unknown: _____ Answer: _____ kilogram(s)

⑤ On the back, write your own multiplicative comparison number story using the data in the chart.

Multiplicative Comparisons

① Record the following statements as equations.

a. 56 is 7 times as many as 8. _____

b. 18 is 6 times as many as 3. _____

c. 63 is 9 times as many as 7. _____

② Use an equation to answer the problems.

a. What number is 5 times as much as 5? _____

b. What number is 4 times as much as 3? _____

③ Use the equation 6 * 8 = 48 to write a multiplicative comparison number story.

_____

_____

_____

④ Maria ran 12 miles last week. Phoebe ran 60 miles last week.
How many times as far did Phoebe run compared to Maria?

12 miles _____

60 miles _____

Equation with unknown: _____

Answer: _____ times as far

⑤ Neve's rope is 2 times as long as Peter's rope. If Peter has 9 feet
of rope, how many feet of rope does Neve have?

a. Equation with unknown: _____

b. Answer: _____ feet of rope

Multiplicative Comparisons

> **Family Note** In this lesson students used comparison statements and equations to represent situations in which one quantity is a number of times as much as another quantity. For example: José saved $5 over the summer. His sister saved 3 times as much. How much money did José's sister save? In this number story students compare the amount of money José saved to the amount his sister saved. Students write the equation $3 * 5 = 15$ to represent this comparison and solve the problem: José's sister saved $15. Because these comparison statements and equations involve multiplication, they are called multiplicative comparisons.

Complete the problems below. Write an equation with a letter for the unknown and solve.

SRB
56-57

① What number is 7 times as much as 9?

Equation with unknown:

_____

Answer: _____

② What number is 5 times as much as 6?

Equation with unknown:

_____

Answer: _____

③ 32 is 4 times as much as what number?

a. Equation with unknown: _____

b. Answer: _____

④ Write an equation to represent this situation and solve.

Ameer worked 3 times as many hours as Simi each week during the summer. If Simi worked 10 hours each week, how many hours did Ameer work each week?

a. Equation with unknown: _____

b. Answer: _____ hours

Practice

⑤ $7,482 - 7,083 =$ _____

⑥ $7,702 - 3,581 =$ _____

⑦ $5,201 - 3,052 =$ _____

⑧ $8,002 - 5,403 =$ _____

Additive Comparison Number Stories

Read the problem. Place the numbers and the unknown into the diagram. Circle the number sentence that matches the problem. Find the answer.

SRB
56

(1) Quinn has 19 more books than Barbara. Barbara has 32 books.
How many books does Quinn have?

   a. $32 - 19 = b$

   b. $19 + b = 32$

   c. $32 + 19 = b$

Answer: _____ books

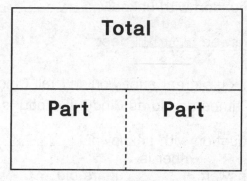

(2) The mail carrier had 121 pieces of mail to deliver on Monday.
On Tuesday she had 54 more pieces to deliver than on Monday.
How many pieces of mail did she deliver on Tuesday?

   a. $121 + 54 = m$

   b. $121 - m = 54$

   c. $121 - 54 = m$

Answer: _____ pieces of mail

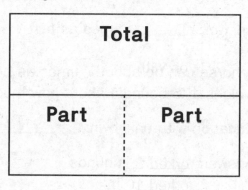

(3) Janice's chapter book has 367 pages. Brian's chapter book has
38 fewer pages. How many pages does Brian's book have?

   a. $38 + 367 = p$

   b. $p - 367 = 38$

   c. $367 - 38 = p$

Answer: _____ pages

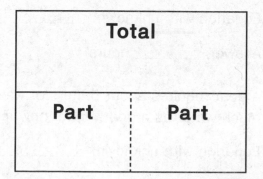

Multiplicative Comparison Number Stories

Write an equation with an unknown and then solve. Use a diagram or drawing as needed.

1. Wallace designed a 57-foot outdoor track for his remote control car. His outdoor track is 3 times as long as his indoor track. What is the measure of his indoor track?

   Equation with unknown: _____

   Answer: _____ feet

2. Cindy's great aunt wouldn't tell Cindy her age, but she did say she was about 5 times as old as Cindy. If Cindy is 14, about how old is her great aunt?

   Equation with unknown: _____

   Answer: _____ years old

3. About how many times as fast is a peregrine falcon diving at 200 miles per hour as a human running at about 25 miles per hour?

   Equation with unknown: _____

   Answer: _____ times as fast

4. A horse can be about 7 times as heavy as an Arctic wolf. If an Arctic wolf weighs 176 pounds, about how much would a horse weigh?

   Equation with unknown: _____

   Answer: _____ pounds

5. A car trip to visit Meri's cousins takes 35 hours, including stops and sleeping. It is 7 times as long as the same trip in a plane. About how many hours is the flight?

   Equation with unknown: _____

   Answer: _____ hours

6. A goat requires about 5 liters of water per day. A dairy cow requires 13 times as much water as a goat. About how much water does a dairy cow require per day?

   Equation with unknown: _____

   Answer: _____ liters

Solving Multiplicative Comparison Number Stories

Make a diagram or drawing and write an equation to represent the situation. Then find the answer.

SRB
56-57

(1) Judith collected 9 marbles. Swen has 6 times as many. How many marbles does Swen have?

Diagram or drawing:

Equation with unknown: _____

Answer: _____ marbles

(2) Sol ran 4 times as many minutes as Jerry. Jerry ran 12 minutes. How many minutes did Sol run?

Diagram or drawing:

Equation with unknown: _____

Answer: _____ minutes

Insert quantities into the number story. Make a diagram and write an equation to represent the story.

(3) Lola picked _____ apples. Eilene picked _____ apples. Eilene picked _____ times as many apples as Lola.

Diagram or drawing:

Equation with unknown: _____

Answer: _____ apples

Practice
Write these numbers in expanded form.

(4) 3,830 _____

(5) 56,037 _____

(6) 800,700 _____

(7) 716,305 _____

77

Identifying Right Triangles

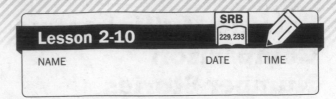
① Trace the triangles on your Geometry Template into the correct column: **Right Triangles** or **Not Right Triangles**. Also draw your own triangles to fit each category.

Right Triangles	Not Right Triangles

② **a.** Use a straightedge. Draw a right triangle.

b. Explain why the shape you drew is a right triangle.

_____

_____

_____

Identifying Triangles

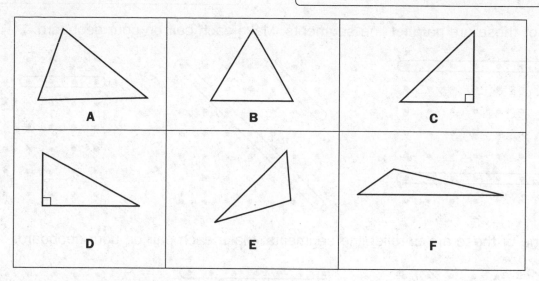

SRB
229, 233

Write the letter or letters that match each statement.

1. Has perpendicular line segments _____

2. Has an obtuse angle _____

3. Has right angles _____

4. Has acute angles _____

5. Has more than one kind of angle _____

6. Has only one kind of angle _____

7. Does NOT have any right angles _____

8. Is a right triangle _____

Practice

9. List all the factors of 12. _____

10. Name the next 4 multiples of 7.   35, _____, _____, _____, _____

79

Parallel Line Segments

(1) All of these are **parallel** line segments. Make each pair on your geoboard.

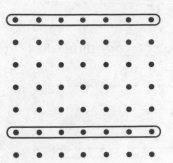 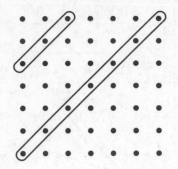 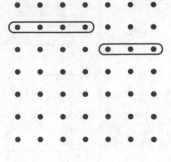

(2) None of these are parallel line segments. Make each pair on your geoboard.

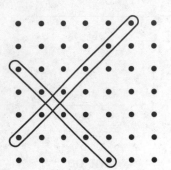 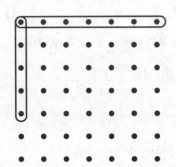 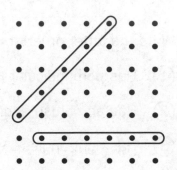

(3) Some of these are parallel line segments. Make each pair on your geoboard. Circle the parallel line segments.

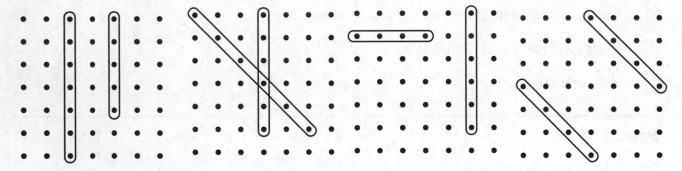

(4) How would you describe parallel line segments to a friend?

_____

_____

_____

_____

(5) Practice making other parallel line segments on your geoboard.

80

Solving Quadrilateral Riddles

Read the clues and then solve.

---

① I have four equal sides.

My opposite sides are parallel.

I have no right angles.

What am I? _____

Draw me.

---

② All of my angles are different.

I have one set of parallel sides.

I have four unequal sides.

What am I? _____

Draw me.

---

③ I have two separate pairs of equal sides.

My equal sides are next to each other.

My four sides cannot be equal.

What am I? _____

Draw me.

---

④ We are parallelograms.

We are three different polygons.

Sometimes we share our names.

What are we?

_____

_____

Draw us.

Drawing Quadrilaterals

(1) A parallelogram is a quadrilateral that has 2 pairs of parallel sides. Draw a parallelogram.

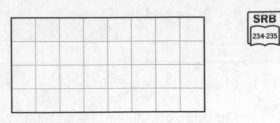

SRB
234-235

(2) Answer each question, drawing pictures on the back of this page to help you.

a. Can a parallelogram have right angles? _____ Explain.

_____

_____

b. Could a quadrilateral have 4 obtuse angles? _____ Explain.

_____

_____

c. Name a quadrilateral that has at least 1 pair of parallel sides.

_____

(3) Draw a quadrilateral that has at least 1 right angle.

(4) Draw a quadrilateral that has 2 separate pairs of equal length sides but is NOT a parallelogram.

This is called a _____.

Practice

(5) 5 * 30 = _____

(6) _____ = 40 * 3

(7) _____ = 80 * 6

(8) 6 * 70 = _____

What's Missing?

Symmetric Pictures

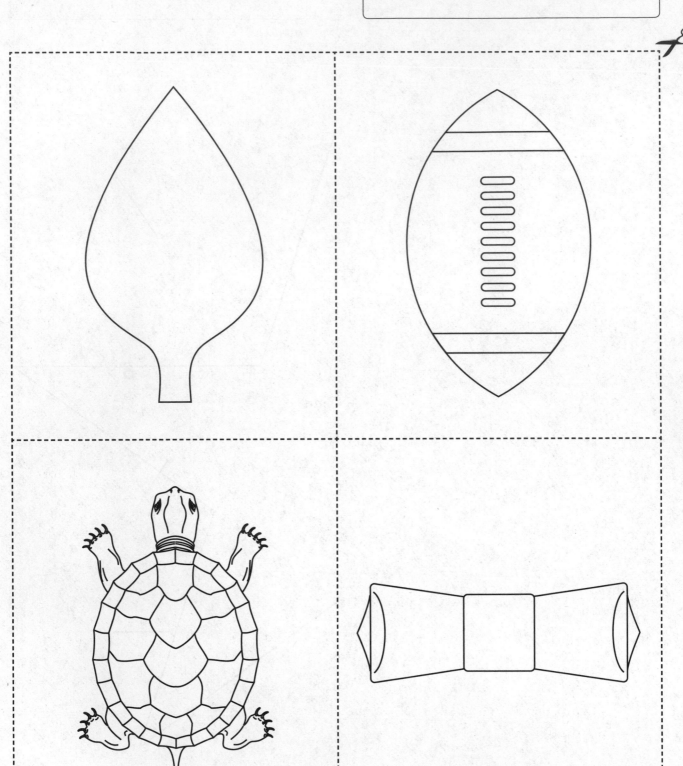

84

Polygons A–E

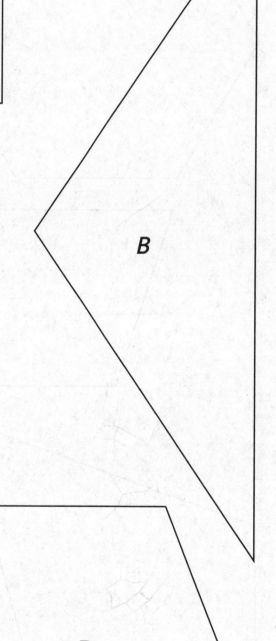

E

C

B

A

D

Polygons F–J

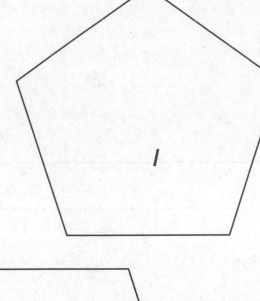

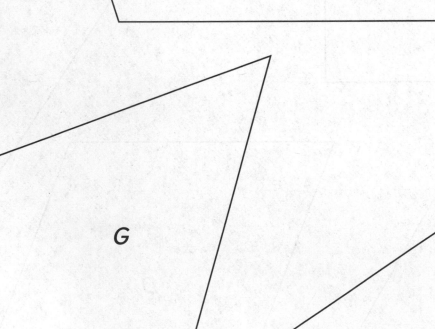

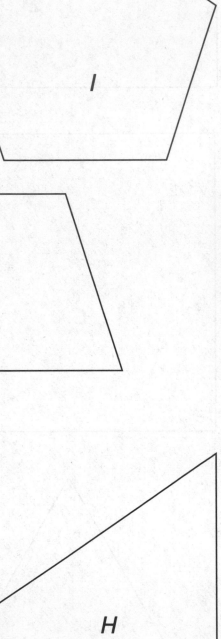

J

I

F

G

H

Line Symmetry in the Alphabet

1 Print the 26 capital letters of the alphabet below.

SRB
238

....

....

2 The capital letter A has a vertical line of symmetry. **A**

The capital letter B has a horizontal line of symmetry. **B**

Use the letters of the alphabet to complete the Venn diagram.

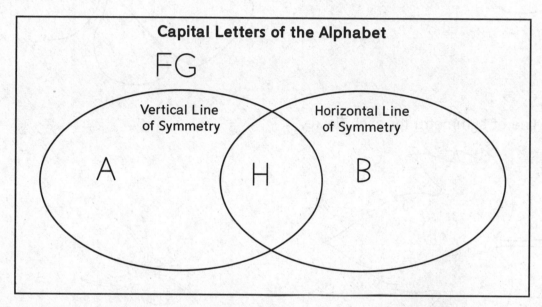

Capital Letters of the Alphabet

FG

Vertical Line of Symmetry   Horizontal Line of Symmetry

A   H   B

3 The word BED has a horizontal line of symmetry. **BED**

The word HIT has a vertical line of symmetry. **HIT**

Use capital letters to list words that have horizontal or vertical line symmetry.

horizontal	vertical
_____	_____
_____	_____

4 Find a word with both horizontal and vertical symmetry. _____

Drawing Lines of Symmetry

① Draw the other half of each picture to make it symmetrical. Use a straightedge to form the line of symmetry.

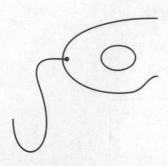 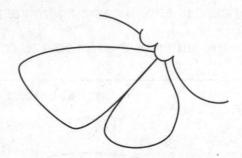

② Draw a line of symmetry for each figure.

③ List four items in your home that are symmetric. Pick one item and draw it below, including at least one line of symmetry.

Item: _____   Item: _____   Drawing:

Item: _____   Item: _____

Practice

④ _____ = 2,767 + 3,254

⑤ 193 + 6,978 = _____

⑥ 7,652 − 5,388 = _____

⑦ _____ = 4,273 − 1,678

88

"What's My Rule?" Number of Sides

① Use square pattern blocks to help you complete the table.

Number of Squares	Number of Sides
1	4
2	8
3	
5	
7	
8	

② Suppose there are 12 squares. Explain how to find the number of sides without counting.

_____

_____

③ Use triangle pattern blocks to help you complete the table.

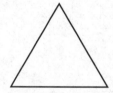

Number of Triangles	Number of Sides
1	3
2	6
	15
	12
	9
	18

④ Suppose there are 30 sides. Explain how to find the number of triangles without counting.

_____

_____

Identifying Patterns

① Complete.

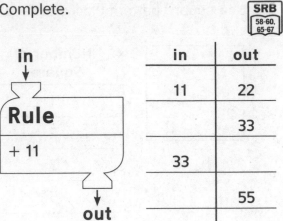

in	out
2	18
	27
4	
	45
6	

Rule

∗ 9

What patterns do you see?

_____

_____

_____

_____

② Complete.

in	out
11	22
	33
33	
	55
55	

Rule

+ 11

What patterns do you see?

_____

_____

_____

_____

③ Study the pattern.

    1   2   3   4   5

a. Draw the next step in the pattern. What patterns do you notice?

_____

_____

b. How many circles will be in the 6th step? _____ In the 10th step? _____

c. How did you figure out how many circles will be in the 10th step?

_____

_____

Practice

④ 800,000 + 90 = _____

⑤ 200,000 + 50,000 + 4 = _____

Fractions and Decimals

Of the different types of numbers that elementary school students are required to understand, fractions can be the hardest. To succeed with fractions, students must build on their understanding of whole numbers, but they also have to understand how fractions differ from whole numbers.

This unit focuses on three big ideas about fractions: recognizing equivalent fractions, comparing fractions, and representing or showing fractions in different ways.

Fraction Equivalence

Equivalence, or equality, is one of the biggest ideas in mathematics. Much of arithmetic, for example, is really just rewriting numbers in equivalent forms. When we ask students to solve $850 + 125$, we are asking for a single number, 975, that is equivalent to $850 + 125$.

Students were introduced to equivalent fractions in third grade. They reasoned about equivalent fractions by thinking about sharing ($\frac{1}{2}$ is a fair share when 1 whole is shared 2 ways), division ($\frac{3}{4}$ is the result of dividing 3 wholes into 4 parts), and measurement ($\frac{1}{2}$ of an inch and $\frac{2}{4}$ of an inch name the same length).

In fourth grade students learn a multiplication rule for making equivalent fractions: *To make a fraction equivalent to a given fraction, multiply the numerator and denominator by the same number (so long as that number is not 0).* For example, to make a fraction equivalent to $\frac{3}{4}$, we can multiply the numerator and denominator by 2: $\frac{3*2}{4*2} = \frac{6}{8}$.

Comparing Fractions

When students compare fractions with the same denominator or numerator, they are also building on previous work by thinking about how familiar things are divided and shared. In third grade students reasoned that $\frac{3}{5}$ is less than $\frac{3}{4}$ because sharing 3 pizzas among 5 people means less pizza for each person compared to sharing the same 3 pizzas among only 4 people.

In fourth grade students continue work with fractions based on visual models and reasoning about sharing. They also learn to use benchmark fractions, such as 0, $\frac{1}{2}$, and 1. Students reason that $\frac{3}{5}$ is more than $\frac{1}{3}$ because $\frac{3}{5}$ is more than $\frac{1}{2}$ and $\frac{1}{3}$ is less than $\frac{1}{2}$.

Representing Fractions

One reason fractions can be hard to understand is that the same number can be shown in so many different ways. For example, $\frac{1}{2}$ can be written as $\frac{2}{4}$, $\frac{3}{6}$, or an infinite number of equivalent fractions. It can also be written as 0.5, 0.50, or an infinite number of other equivalent decimals. And the number $\frac{1}{2}$ can represent half the amount of different whole units.

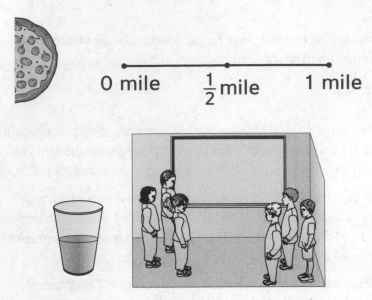

0 mile $\frac{1}{2}$ mile   1 mile

Different representations of $\frac{1}{2}$

In this unit students work with fractions and decimals represented in many different ways, using concrete objects like fraction circle pieces, base-10 blocks, and strips of paper; hundred grids and number lines; drawings of circles and rectangles; and rulers and other measuring tools. For example, they might show that $\frac{3}{4}$ is greater than $\frac{5}{8}$ using fraction circle pieces, folding paper strips, pointing to marks on a ruler, or finding equivalent fractions with a common denominator or common numerator.

In this way students build a network of ideas that help them develop a solid conceptual understanding of fractions and decimals, which will support work in later units and grades that is focused on more formal rules and procedures.

Please keep this Family Letter for reference as your child works through Unit 3.

Vocabulary

Important terms in Unit 3:

benchmark A count or measure that can be used to evaluate the reasonableness of other counts, measures, or estimates. For example, a benchmark for land area is a football field, which is about 1 acre.

centimeter A metric unit of length equivalent to 10 millimeters, $\frac{1}{10}$ of a decimeter, and $\frac{1}{100}$ of a meter.

common denominator A nonzero number that is a multiple of the denominators of two or more fractions. For example, the fractions $\frac{1}{2}$ and $\frac{2}{3}$ have common denominators 6, 12, 18, and so on.

common numerator A nonzero number that is a multiple of the numerators of two or more fractions. For example, the fractions $\frac{3}{4}$ and $\frac{4}{5}$ have common numerators 12, 24, 36, and so on.

denominator The nonzero digit b in a fraction $\frac{a}{b}$. In a part-whole fraction, the denominator is the number of equal parts into which the whole has been divided.

equivalent Equal in value, but possibly represented in a different form. For example, $\frac{1}{2}$, 0.5, and 50% are all equivalent.

Equivalent Fractions Rule A rule stating that if the numerator and denominator of a fraction are multiplied by the same nonzero number, the result is a fraction that is equivalent to the original fraction. This rule can be represented as:
$$\frac{a}{b} = \frac{(n*a)}{(n*b)}.$$

hundredth A single part out of one hundred equal parts that form a whole.

interval The points and their coordinates on a segment of a number line. The interval between 0 and 1 on a number line is the *unit interval*.

meter The basic metric unit of *length* from which other metric units of length are derived. One meter is equal to 10 decimeters, 100 centimeters, or 1,000 millimeters. A meter is a little longer than a yard.

metric The measurement system used in most countries and by virtually all scientists around the world. Units in the metric system are related by powers of 10.

millimeter A metric unit of length equal to $\frac{1}{10}$ of a centimeter or $\frac{1}{1,000}$ of a meter.

numerator The digit a in a fraction $\frac{a}{b}$. In a part-whole fraction in which the whole is divided into a number of equal parts, the numerator is the number of equal parts being considered.

reasoning An explanation or justification for how to solve a problem or answer a question.

representation Something that shows, symbolizes, or stands for something else. For example, numbers can be represented using base-10 blocks, spoken words, or written numerals.

strategy A general approach to solving a problem or answering a question.

tenth A single part out of ten equal parts that form a whole.

unit A label used to put a number in context. In measuring length, for example, inches and centimeters are units. In a problem about 5 apples, the unit is apples. In *Everyday Mathematics* students keep track of units in *unit boxes*.

whole An entire object, collection of objects, or quantity being considered in a problem situation; 100%.

Do-Anytime Activities

To work with your child on concepts taught in this unit, try these activities:

1. Have your child look for everyday uses of fractions in grocery stores, shoe sizes, cookbooks, measuring cups, and statistics in newspapers and on television.

2. Encourage your child to express fractions, quantities, and measures, such as a quarter of an hour, a quart of orange juice, or a quarter cup of milk.

3. Encourage your child to incorporate terms such as *whole, halves, thirds*, and *fourths* into his or her everyday vocabulary.

Building Skills through Games

In this unit your child will play the following games to develop his or her understanding of fractions and decimals. For detailed instructions, see the *Student Reference Book.*

Fraction Match See *Student Reference Book,* page 263. This game is for 2 to 4 players and requires one set of fraction cards. The game develops skill in naming equivalent fractions.

Fraction Top-It See *Student Reference Book,* page 265. This is a game for 2 to 4 players and requires one set of fraction cards. The game develops skill in comparing fractions.

Decimal Top-It See *Student Reference Book,* page 253. This is a game for 2 to 4 players and requires one set of number cards and a gameboard. The game develops skill in comparing decimals.

As You Help Your Child with Homework

As your child brings assignments home, you may want to go over instructions together, clarifying them as necessary. The answers listed below will guide you through the Home Links for this unit.

Home Link 3-1

1. $1\frac{1}{4}$, or $\frac{5}{4}$ pizzas

One way:

Another way:

3. 14, 21, 28, 35   **5.** 2, 3

Home Link 3-2

1.   3.

6. 25, 30, 35, 40   **8.** 4, 6, 8, 12, 16, 24, 48

Home Link 3-3

1. **a.** ≠   **b.** =   **c.** ≠   **d.** =   **e.** =

3. *c, d,* and *e* should be circled

5. 8,033   7. 288

Home Link 3-4

1. 12   **3.** 6   **5.** 2   **7.** 12   **9.** 3

11. Sample answers: $\frac{2}{4}, \frac{4}{8}, \frac{5}{10}, \frac{6}{12}$

13. 1 and 30, 2 and 15, 3 and 10, 5 and 6

Home Link 3-5

1. **a.** $\frac{3}{4}$ **b.** Answers vary.

3. They will get the same amount; $\frac{3}{4}$ and $\frac{6}{8}$ are equivalent fractions.

6. 1 and 75, 3 and 25, 5 and 15

Home Link 3-6

1. They have both read the same amount; Sample answer: $\frac{3}{4}$ is the same as $\frac{6}{8}$.

2. Heather; $\frac{5}{8} > \frac{5}{10}$, or $\frac{5}{10} < \frac{5}{8}$

3. Howard's; Sample answers: Jermaine's team won less than half of its games, and Howard's team won more than half; $\frac{2}{5}$ is the same as $\frac{4}{10}$, and $\frac{7}{10} > \frac{4}{10}$.

5. F   7. T

Home Link 3-7

1. $\frac{2}{6}, \frac{4}{6}, \frac{5}{6}$

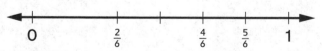

3. $\frac{1}{6}, \frac{4}{10}, \frac{1}{2}, \frac{7}{12}, \frac{2}{3}$

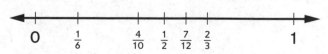

5. 10,121   7. 1,329

Home Link 3-8

1.

Number in Words	Fraction	Decimal
one-tenth	$\frac{1}{10}$	0.1
four-tenths	$\frac{4}{10}$	0.4
eight-tenths	$\frac{8}{10}$	0.8
nine-tenths	$\frac{9}{10}$	0.9
two-tenths	$\frac{2}{10}$	0.2
seven-tenths	$\frac{7}{10}$	0.7

3. 0.5 **5.** 0.9

7. 1, 2, 4, 5, 10, 20, 25, 50, 100

9. 1 and 42, 2 and 21, 3 and 14, 6 and 7

Home Link 3-9

1. $\frac{30}{100}$; 0.30 **3.** $\frac{65}{100}$; 0.65

5. 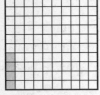 **7.** 3, 9

Home Link 3-10

1. 57; 5; 7 **3.** 4; 0; 4 **5.** 8.4 **7.** 0.05

9. 0.04, 0.05, 0.06, 0.07, 0.08, 0.09

11. 46,000

Home Link 3-11

1. 0.02; 0.03; 0.04; 0.05; 0.06; 0.07

3–6.

7. a. > **b.** < **c.** > **9.** 13,931 **11.** 1,569

Home Link 3-12

1. Answers vary. **3.** 100; 1.8; 2,360; 572; 65

5. 1 and 60, 2 and 30, 3 and 20, 4 and 15, 5 and 12, 6 and 10

Home Link 3-13

1. < **3.** > **5.** > **7.** =

9. hundredths; 0.09 **11.** 6.59, 6.60, 6.61

13. 4.4 **15.** 8.1

17. 56,230 **19.** 15,379

Solving a Proportional Reasoning Problem

Solve. Show your work.

1 Three students in Mrs. Orr's science class are asked to share 5 feet of string for a project involving pulleys. How much string does each student get?

_____ feet of string

2 Seven students in Mr. Flynn's class need string for the same project. How much string should Mrs. Orr send to Mr. Flynn so that each of his students gets the same amount of string as her students? Explain your reasoning.

_____ feet of string

Solving Equal-Sharing Number Stories

Use drawings to help you solve the following problems. Try to solve each problem in more than one way. Show your work.

SRB
124-125,
156-157

① Two friends are sharing 5 banana muffins. How much muffin will each person get, if each gets the same amount?

_____ banana muffins

One way: Another way:

② Mr. Wells has 8 chicken pot pies to share among 6 people. If each person gets the same-size serving, how much chicken pot pie will each person get?

_____ chicken pot pies

One way:

Another way:

Sharing Equally

Use drawings to help you solve the problems. Solve each problem in more than one way. Show your work.

SRB
124-125, 156-157

① Four friends shared 5 pizzas equally. How much pizza did each friend get?

_____ pizzas

One way:

Another way:

② Five kittens are sharing 6 cups of milk equally. How much milk does each kitten get?

_____ cups of milk

One way:

Another way:

Practice

③ Name the next 4 multiples of 7. 7, _____, _____, _____, _____

④ List all the factors of 18. _____

⑤ List all the factors of 18 that are prime. _____

⑥ List all the factor pairs of 40.

_____ and _____ ; _____ and _____ ;

_____ and _____ ; _____ and _____

99

Fraction Fill

Cover each region with fraction circle pieces. Use only one color at a time.
Try all of the colors. Record your work on journal page 71.

Whole

red circle

① $\frac{2}{2}$

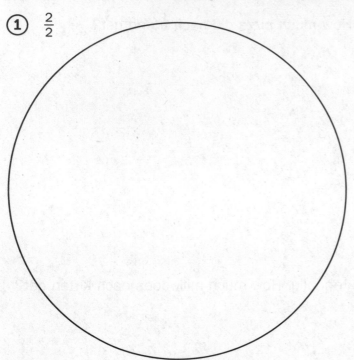

② $\frac{1}{3}$

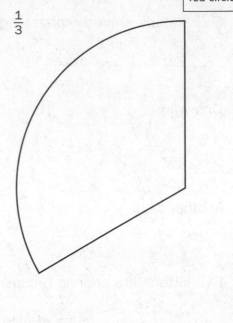

③ $\frac{2}{3}$

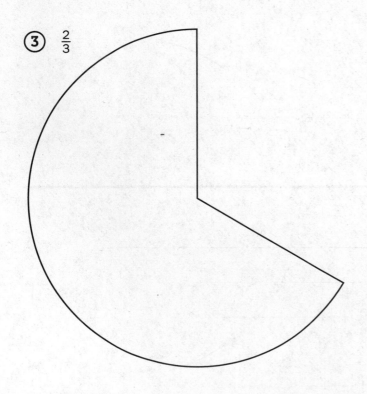

④ $\frac{1}{4}$

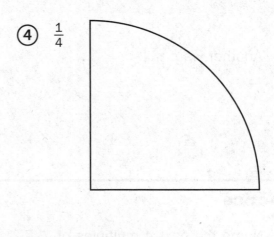

Fraction Fill (continued)

Whole
red circle

⑤ $\frac{3}{4}$

⑥ $\frac{1}{5}$

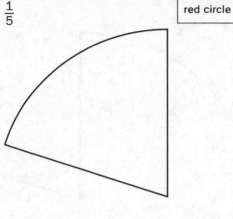

⑦ $\frac{2}{5}$

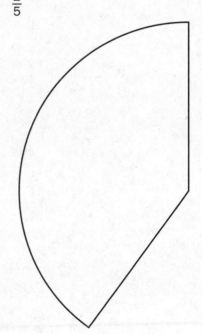

⑧ $\frac{3}{5}$

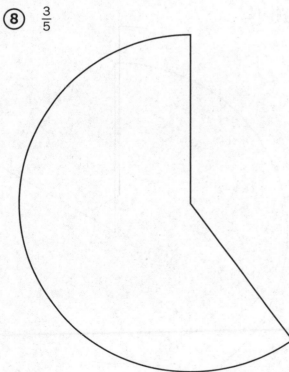

101

Fraction Fill (continued)

9 $\frac{4}{5}$

10 $\frac{1}{6}$

Whole
red circle

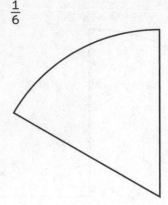

11 $\frac{5}{6}$

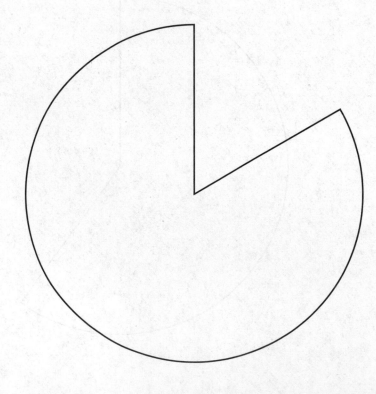

Exploring Fraction Circles

① 1 whole circle equals ____ oranges.

② ____ pinks equal 1 whole circle.

③ 1 yellow equals ____ dark blues.

④ ____ purples equal 1 dark green.

⑤ ____ light blues equal 1 orange.

Use <, >, or =.

⑥ 1 pink ____ 1 orange

⑦ 1 dark green ____ 1 yellow

⑧ 1 red ____ 4 yellows

⑨ Shelby covered a red circle with 2 yellows and 4 dark blues. She said, "One dark blue is $\frac{1}{6}$ of a red circle." Do you agree or disagree? Explain.

Whole
red circle

_____

_____

_____

⑩ Ava said purple is $\frac{1}{10}$. Tyler said purple is $\frac{1}{4}$. Mrs. Robey looked at the whole each student had on their desks and said, "You are both correct." Explain how this could be true.

Modeling Fraction Equivalencies

① Explain why a hexagon pattern block is useful for modeling equivalencies of fractions with denominators of 2, 3, and 6.

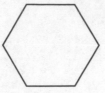 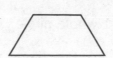

_____

_____

_____

_____

② Study the clock face. Which denominators can be modeled on the clock face?

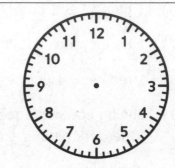

_____

_____

Explain your answer.

_____

_____

③ Using the denominators from Problem 2, name the fraction represented on each clock face in as many different ways as you can.

a.

_____

b.

_____

c.

_____

d.

_____

e.

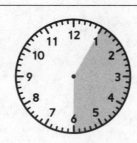

_____

f.

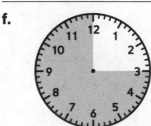

_____

Fraction Circles

(1) Divide into 4 equal parts. Shade $\frac{1}{4}$.

(2) Divide into 8 equal parts. Shade $\frac{2}{8}$.

SRB
136

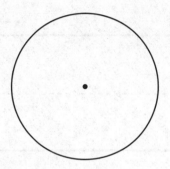

(3) Divide into 12 equal parts. Shade $\frac{3}{12}$.

(4) Create your own. Divide into equal parts and shade a portion. Record the amount you shaded.

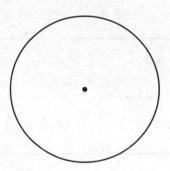

(5) What patterns do you notice in Problems 1 through 3?

_____

_____

Practice

(6) List the next 4 multiples of 5. 20, _____, _____, _____, _____

(7) List all the factors of 48. _____

(8) List the factors of 48 that are composite. _____

Fraction Number Lines

Lesson 3-3

NAME DATE TIME

Cut along the dashed lines.

1 Whole

0 1

Halves

$\frac{0}{2}$ $\frac{1}{2}$ $\frac{2}{2}$

Fifths

Tenths

Thirds

Sixths

Twelfths

Finding Equivalent Fractions

Use fraction circles to find equivalent fractions. Fill in the missing numerators to complete the equivalent fractions.

(1) Cover $\frac{1}{2}$ of the circle with fourths.

whole
red circle

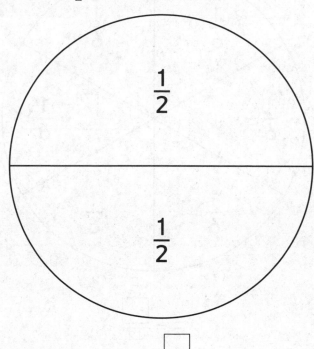

$$\frac{1}{2} = \frac{\square}{4}$$

(2) Cover $\frac{1}{4}$ of the circle with eighths.

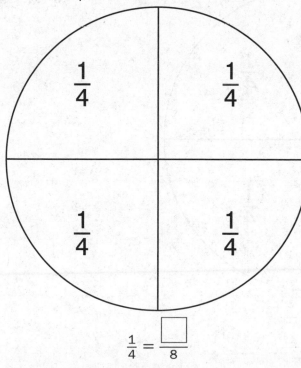

$$\frac{1}{4} = \frac{\square}{8}$$

(3) Cover $\frac{1}{2}$ of the circle with sixths.

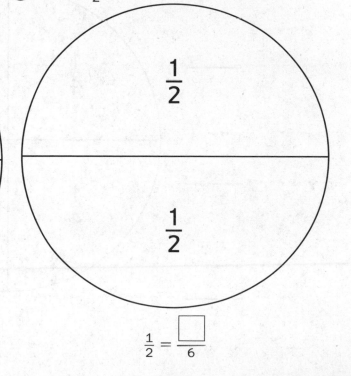

$$\frac{1}{2} = \frac{\square}{6}$$

Finding Equivalent Fractions (continued)

④ Cover $\frac{2}{5}$ of the circle with tenths.

⑤ Cover $\frac{4}{6}$ of the circle with twelfths.

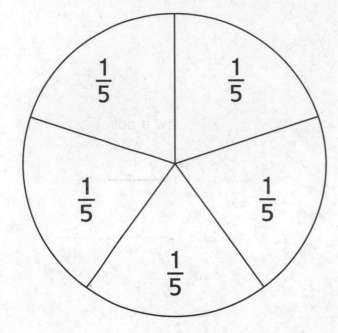

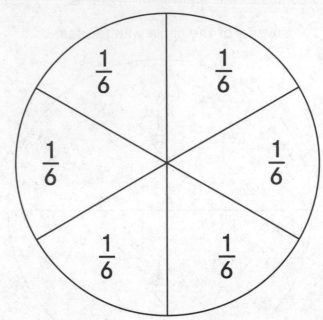

$$\frac{2}{5} = \frac{\square}{10}$$

$$\frac{4}{6} = \frac{\square}{12}$$

⑥ Cover $\frac{2}{4}$ of the circle with twelfths.

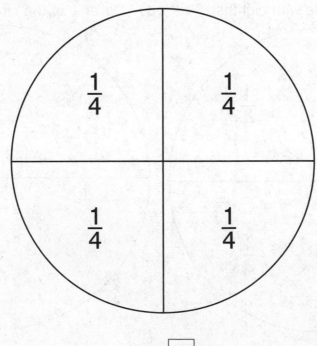

$$\frac{2}{4} = \frac{\square}{12}$$

Identifying Equivalent Fractions on Number Lines

Follow the directions below.

(1) Look at point Y on the number line.

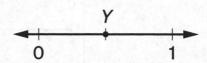

Look at the number lines below. Circle the number lines that show a point equal to the number represented by Y.

a.

b.

c.

d.

e.

(2) Look at point Z on the number line.

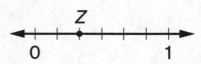

Look at the number lines below. Circle the number lines that show a point equal to the number shown by Z.

a.

b.

c.

d.

e.

109

Finding Equivalent Fractions

Use the number lines to help you answer the following questions.

(1) Fill in the blank with = or ≠.

a. $\frac{2}{3}$ _____ $\frac{1}{3}$

b. $\frac{2}{6}$ _____ $\frac{1}{3}$

c. $\frac{2}{6}$ _____ $\frac{2}{5}$

d. $\frac{1}{5}$ _____ $\frac{2}{10}$

e. $\frac{2}{12}$ _____ $\frac{1}{6}$

(The number lines show 0 to 1 divided into halves, thirds, fifths, sixths, tenths, and twelfths.)

(2) Fill in the missing numbers.

a. $\frac{1}{5} = \frac{\square}{10}$

b. $\frac{4}{12} = \frac{\square}{3}$

c. $\frac{5}{10} = \frac{\square}{2}$

d. $\frac{3}{6} = \frac{\square}{12}$

e. $\frac{4}{6} = \frac{\square}{3}$

(3) Circle the number sentences that are NOT true.

a. $\frac{3}{12} = \frac{1}{4}$

b. $\frac{1}{2} = \frac{5}{10}$

c. $\frac{2}{6} = \frac{2}{5}$

d. $\frac{7}{10} = \frac{4}{6}$

e. $\frac{9}{10} = \frac{11}{12}$

Practice

Solve using U.S. traditional addition or subtraction.

(4) _____ = 989 + 657

(5) 3,314 + 4,719 = _____

(6) 5,887 − 3,598 = _____

(7) _____ = 2,004 − 1,716

An Equivalent Fractions Rule

Margot says the value of a fraction does not change if you do the same thing to the numerator and denominator. She says that she added 2 to the numerator and denominator in $\frac{1}{4}$ and got $\frac{3}{6}$.

$$\frac{(1 + 2)}{(4 + 2)} = \frac{3}{6}$$

She says that therefore $\frac{1}{4} = \frac{3}{6}$. How can you tell or show Margot that she is wrong?

✁ -

An Equivalent Fractions Rule

Lesson 3-4

NAME   DATE   TIME

Margot says the value of a fraction does not change if you do the same thing to the numerator and denominator. She says that she added 2 to the numerator and denominator in $\frac{1}{4}$ and got $\frac{3}{6}$.

$$\frac{(1 + 2)}{(4 + 2)} = \frac{3}{6}$$

She says that therefore $\frac{1}{4} = \frac{3}{6}$. How can you tell or show Margot that she is wrong?

Exploring Fractions of Circles

Divide each circle into equal parts and color as directed.

(1) Divide into 2 equal parts.
Color $\frac{1}{2}$ pink.

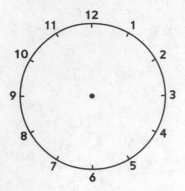

(2) Divide into 3 equal parts.
Color $\frac{2}{3}$ orange.

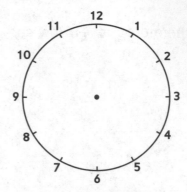

(3) Divide into 6 equal parts.
Color $\frac{2}{6}$ light blue.

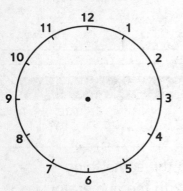

(4) Divide into 6 equal parts.
Color $\frac{3}{6}$ light blue and $\frac{1}{2}$ pink.

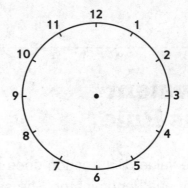

(5) Divide into 6 equal parts.
Color $\frac{2}{6}$ light blue and $\frac{1}{3}$ orange.

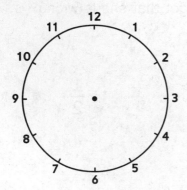

(6) Divide into 12 equal parts.
Color $\frac{1}{4}$ yellow and $\frac{3}{12}$ green.

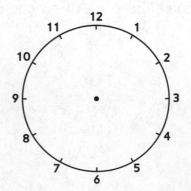

112

Modeling Fraction Equivalencies

Study the example below. Match each fraction in the left column with an equivalent fraction in the right column. Then fill in the boxes on the left showing how you changed each fraction to get an equivalent fraction.

Example:

$$\frac{2 \boxed{* 3}}{3 \boxed{* 3}} \qquad \frac{4}{10}$$

$$\frac{2 \boxed{* 2}}{5 \boxed{* 2}} \qquad \frac{6}{9}$$

$$\frac{1 \boxed{* 2}}{4 \boxed{* 2}} \qquad \frac{2}{8}$$

①

$$\frac{1 \boxed{}}{3 \boxed{}} \qquad \frac{6}{12}$$

$$\frac{3 \boxed{}}{5 \boxed{}} \qquad \frac{4}{12}$$

$$\frac{3 \boxed{}}{6 \boxed{}} \qquad \frac{6}{10}$$

②

$$\frac{1 \boxed{}}{2 \boxed{}} \qquad \frac{9}{12}$$

$$\frac{3 \boxed{}}{4 \boxed{}} \qquad \frac{4}{6}$$

$$\frac{2 \boxed{}}{3 \boxed{}} \qquad \frac{4}{8}$$

③ Now use division.

$$\frac{4 \boxed{}}{8 \boxed{}} \qquad \frac{1}{6}$$

$$\frac{2 \boxed{}}{12 \boxed{}} \qquad \frac{4}{6}$$

$$\frac{8 \boxed{}}{12 \boxed{}} \qquad \frac{1}{2}$$

Finding Equivalent Fractions

Family Note Today students learned about an **Equivalent Fractions Rule**, which can be used to rename any fraction as an equivalent fraction. The rule for multiplication states that if the numerator and denominator are multiplied by the same nonzero number, the result is a fraction that is equivalent to the original fraction.

For example, the fraction $\frac{1}{2}$ can be renamed as an infinite number of equivalent fractions. When you multiply the numerator 1 by 5, the result is 5. When you multiply the denominator 2 by 5, the result is 10.

$$\frac{1 \times 5}{2 \times 5} = \frac{5}{10}$$

This results in the number sentence $\frac{1}{2} = \frac{5}{10}$. If you multiplied both the numerator and denominator in $\frac{1}{2}$ by 3, the result would be $\frac{3}{6}$, which is also equal to $\frac{1}{2}$.

Fill in the boxes to complete the equivalent fractions.

Example: $\frac{1}{2} = \dfrac{3}{\boxed{6}}$

① $\frac{1}{2} = \dfrac{6}{\boxed{}}$ ② $\frac{1}{4} = \dfrac{3}{\boxed{}}$ ③ $\frac{1}{3} = \dfrac{2}{\boxed{}}$ ④ $\frac{2}{3} = \dfrac{8}{\boxed{}}$ ⑤ $\frac{1}{5} = \dfrac{\boxed{}}{10}$

⑥ $\frac{2}{5} = \dfrac{\boxed{}}{10}$ ⑦ $\frac{3}{4} = \dfrac{9}{\boxed{}}$ ⑧ $\frac{5}{6} = \dfrac{10}{\boxed{}}$ ⑨ $\dfrac{2}{\boxed{}} = \dfrac{6}{9}$ ⑩ $\dfrac{4}{\boxed{}} = \dfrac{8}{12}$

⑪ Name 3 equivalent fractions for $\frac{1}{2}$. _____

Practice

⑫ List all the factors of 56. _____

⑬ Write the factor pairs for 30.

_____ and _____, _____ and _____, _____ and _____,

_____ and _____

⑭ Is 30 prime or composite? _____

114

Veggie Pizzas

Lesson 3-5

NAME DATE TIME

In a fourth-grade class small groups of students went on different field trips. The cafeteria prepared 17 veggie pizzas for the students. Since each group had a different number of students, they were given different numbers of pizzas as shown in the diagram below. All the pizzas were the same size.

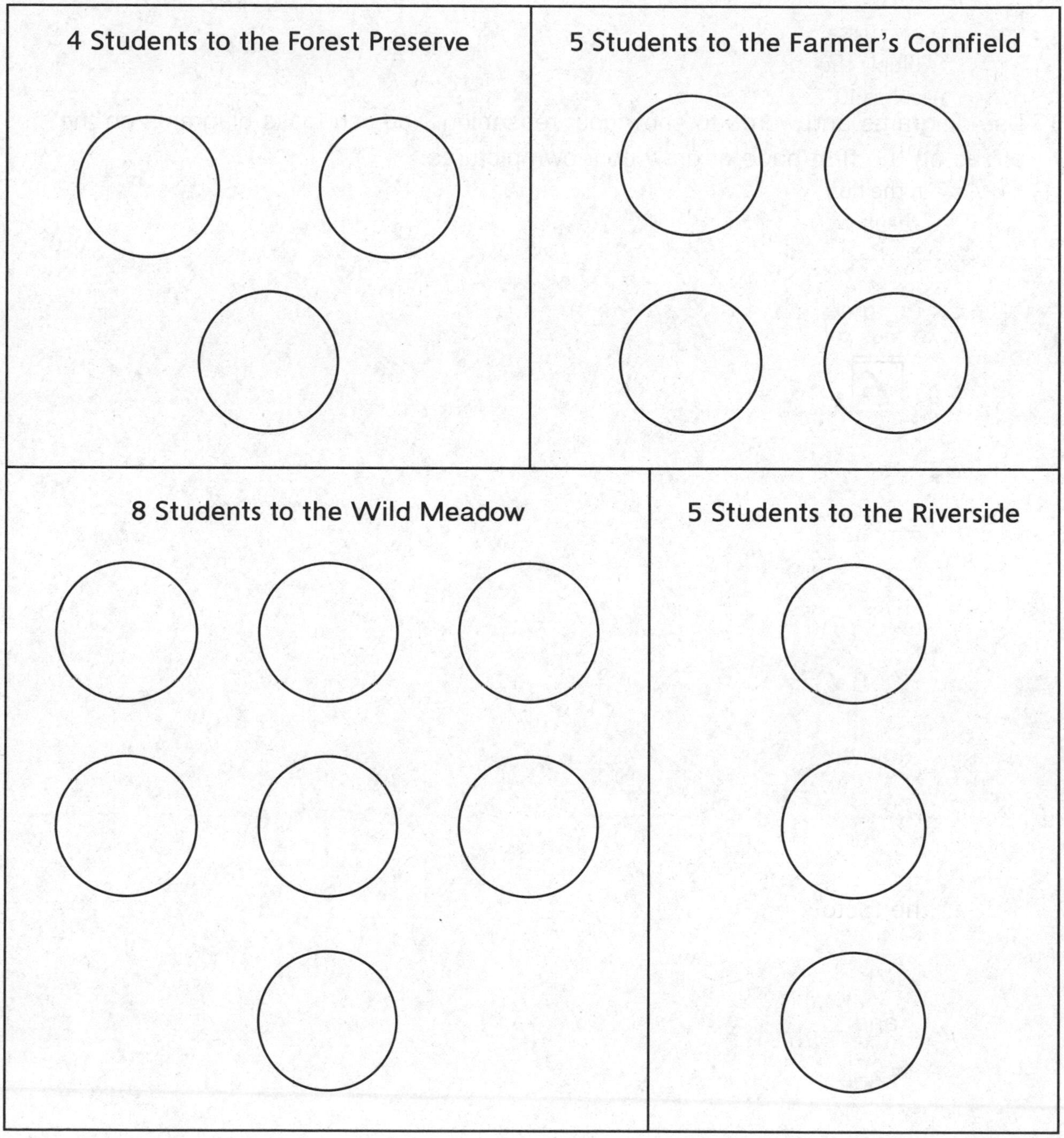

4 Students to the Forest Preserve
5 Students to the Farmer's Cornfield
8 Students to the Wild Meadow
5 Students to the Riverside

Veggie Pizzas (continued)

① In which group did each student have the greatest amount of veggie pizza?

② Use diagrams and words to show your reasoning. You can make diagrams on the pizzas on the first page or draw your own pictures.

Sharing Veggie Pizza

① Karen and her 3 friends want to share 3 small veggie pizzas equally. Karen tried to figure out how much pizza each of the 4 children would get. She drew this picture and wrote two answers.

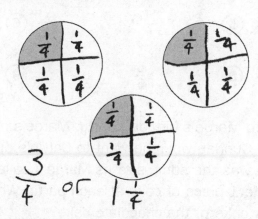

a. Which of Karen's answers is correct? _____

b. Draw on Karen's diagram to make it clear how the pizza should be distributed among the 4 children.

② Erin and her 7 friends want to share 6 small veggie pizzas equally.

How much pizza will each of the 8 children get? _____

③ Who will get more pizza, Karen or Erin? _____

Explain or show how you know.

Practice

④ List all the factors of 50. _____

⑤ Is 50 prime or composite? _____

⑥ Write the factor pairs for 75.

_____ and _____

_____ and _____

_____ and _____

Comparing Fractions: Same Numerator or Denominator

Fill in the blanks with > or <.

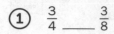

SRB
145

(1) $\dfrac{3}{4}$ ___ $\dfrac{3}{8}$

(2) $\dfrac{7}{10}$ ___ $\dfrac{9}{10}$

(3) $\dfrac{2}{5}$ ___ $\dfrac{2}{12}$

(4) $\dfrac{1}{2}$ ___ $\dfrac{1}{3}$

(5) $\dfrac{2}{6}$ ___ $\dfrac{4}{6}$

(6) $\dfrac{5}{12}$ ___ $\dfrac{8}{12}$

(7) $\dfrac{5}{6}$ ___ $\dfrac{5}{10}$

(8) $\dfrac{6}{12}$ ___ $\dfrac{6}{8}$

(9) $\dfrac{3}{4}$ ___ $\dfrac{3}{5}$

(10) On Saturday Eric went to Marge's birthday party. Marge's mother cut a yellow cake into 8 equal pieces. On Sunday when he went to Calvin's birthday party, Eric noticed that the chocolate cake was the same size as Marge's cake, but it was cut into 12 equal pieces. Eric ate 1 piece of cake at each party. Which piece was larger, the one from the yellow cake or the chocolate cake? _____ piece

    a. Color the amount of each cake that Eric ate.

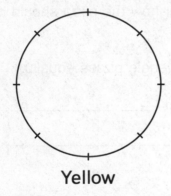

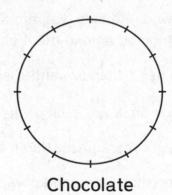

Yellow   Chocolate

    b. Write a number sentence to compare the fractions of the cakes that Eric ate.

_____

Comparing Fractions in Number Stories

For each problem fill in the whole box, answer the question, and write a fraction comparison using <, >, or =.

Draw pictures to show how you found your answers.

(1) Each girl had a ball of yarn. Lisette used $\frac{5}{6}$ of her yarn in one week. Carmen used $\frac{9}{12}$ of her yarn that same week. Who used more yarn that week, or did they use the same amount?

Whole

Answer: _____ Number model: _____

(2) When the two boys returned from hiking on different trails, Marcus said he had hiked $\frac{6}{8}$ of a mile. David said he had hiked $\frac{3}{4}$ of a mile. Who hiked farther, or did they hike the same distance?

Whole

Answer: _____ Number model: _____

(3) Sarah and her sister spent 1 hour at the pool. Sarah spent $\frac{2}{3}$ of the time swimming laps. Her sister spent $\frac{7}{12}$ of the time swimming laps. Who spent more time swimming laps, or did they spend the same amount of time?

Whole

Answer: _____ Number model: _____

Solving Fraction Comparison Number Stories

Solve the problems below.

SRB
145-146

(1) Tenisha and Christa were each reading the same book. Tenisha said she was $\frac{3}{4}$ of the way done with it, and Christa said she was $\frac{6}{8}$ of the way finished.

Who has read more, or have they read the same amount? _____

_____

How do you know? _____

_____

(2) Heather and Jerry each bought an ice cream bar. Although the bars were the same size, they were different flavors. Heather ate $\frac{5}{8}$ of her ice cream bar, and Jerry ate $\frac{5}{10}$ of his.

Who ate more, or did they eat the same amount? _____

Write a number sentence to show this. _____

(3) Howard's baseball team won $\frac{7}{10}$ of its games. Jermaine's team won $\frac{2}{5}$ of its games. They both played the same number of games.

Whose team won more games, or did they win the same amount? _____

How do you know? _____

_____

(4) Write your own fraction number story. Ask someone at home to solve it.

_____

_____

_____

Practice

Write T for true or F for false.

(5) $1,286 + 2,286 = 3,752$ _____

(6) $9,907 - 9,709 = 200$ _____

(7) $2,641 + 4,359 = 2,359 + 4,641$ _____

(8) $2,345 - 198 = 2,969 - 822$ _____

Sorting Fractions

Cut out the cards. Sort the cards into groups according to the fractions shown on them and tape them onto a separate sheet of paper. Next to each group, write why you chose to put the cards into the group.

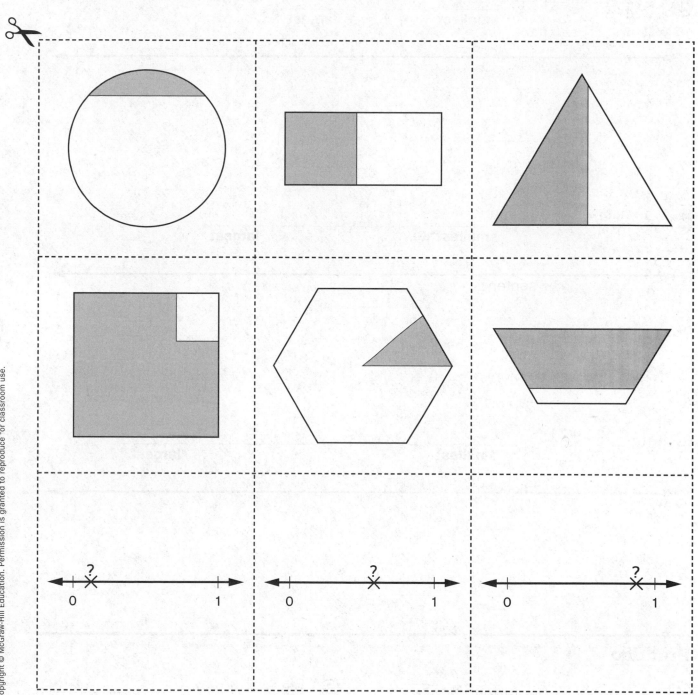

121

Comparing and Ordering Fractions

Write the fractions from smallest to largest, and then justify your conclusions by placing the numbers in the correct places on the number lines.

SRB
135,
147-148

① $\frac{5}{6}$, $\frac{2}{6}$, $\frac{4}{6}$

_____ _____ _____
smallest   largest

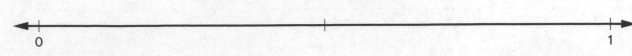

0   1

② $\frac{3}{5}$, $\frac{9}{10}$, $\frac{1}{4}$, $\frac{5}{12}$

_____ _____ _____ _____
smallest   largest

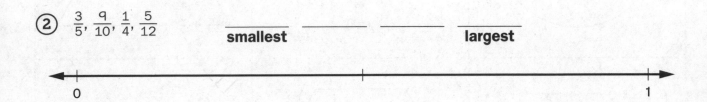

0   1

③ $\frac{7}{12}$, $\frac{1}{2}$, $\frac{2}{3}$, $\frac{4}{10}$, $\frac{1}{6}$

_____ _____ _____ _____ _____
smallest   largest

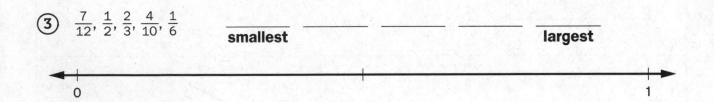

0   1

Practice

④ _____ = 5,494 + 3,769   ⑤ 5,853 + 4,268 = _____

⑥ _____ = 8,210 − 6,654   ⑦ 7,235 − 5,906 = _____

Exploring Hundredths with a Fraction/Decimal Wheel

Cut out the two circles below.

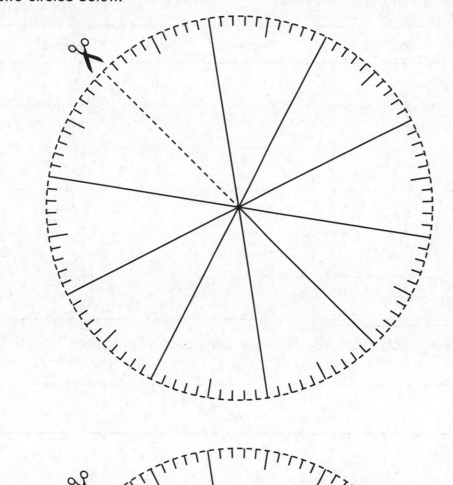

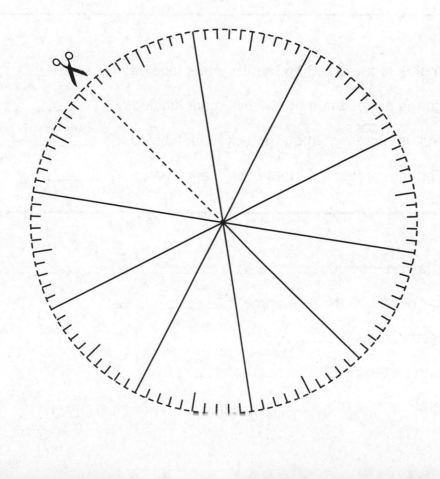

123

Names for Fractions and Decimals

(1) Fill in the blanks in the table below.

Number in Words	Fraction	Decimal
one-tenth		
four-tenths		
	$\frac{8}{10}$	
		0.9
	$\frac{2}{10}$	
seven-tenths		

(2) Name two ways you might see decimals used outside of school.

_____

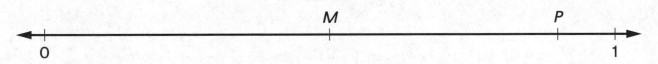

(3) What decimal is represented by the tick mark labeled *M*? _____

(4) What fraction is represented by the tick mark labeled *M*? _____

(5) What decimal is represented by the tick mark labeled *P*? _____

(6) What fraction is represented by the tick mark labeled *P*? _____

Practice

(7) List all the factors of 100. _____

(8) List the factors of 100 that are prime. _____

(9) Write the factor pairs for 42.

_____ and _____   _____ and _____

_____ and _____   _____ and _____

Math Message

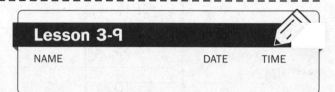

Write the following amounts using a dollar sign and a decimal point:

① 3 dollar bills, 5 dimes, and 1 penny _____

② 3 dimes and 6 pennies _____

③ 2 dollar bills and 7 dimes _____

④ 9 pennies _____

- -

Math Message

Lesson 3-9

NAME   DATE   TIME

Write the following amounts using a dollar sign and a decimal point:

① 3 dollar bills, 5 dimes, and 1 penny _____

② 3 dimes and 6 pennies _____

③ 2 dollar bills and 7 dimes _____

④ 9 pennies _____

- -

Math Message

Lesson 3-9

NAME   DATE   TIME

Write the following amounts using a dollar sign and a decimal point:

① 3 dollar bills, 5 dimes, and 1 penny _____

② 3 dimes and 6 pennies _____

③ 2 dollar bills and 7 dimes _____

④ 9 pennies _____

Money and Decimals

Use only $1 bills , dimes , and pennies .

(1) Use as few bills and coins as possible to show each amount below. Record your work.

Amount	$1 Bills	Dimes	Pennies
$1.26	1	2	6
$1.11			
$2.35			
$3.40			
$2.06			
$0.96			
$0.70			
$0.03			

(2) Describe any patterns you see in the table.

_____

_____

_____

(3) You can use $1 bills, dimes, and pennies to make any amount of money.
Why do you think we have nickels, quarters, and half-dollars?

_____

_____

_____

The Whole

Use base-10 blocks to help you solve the following problems.

① If ▯ is the whole, then what is ▱? _____ What is ▦? _____

② If ▦▦ is the whole, then what is ▯? _____ What is ▱? _____

③ If ▯▯▯▯▯ is the whole, then what is ▱? _____ What is ▦? _____

④ If ▱ ▱ ▱ ▱ ▱ is 0.01, then what is the whole? _____

⑤ If ▯▯▯ is 0.1, then what is the whole? _____

⑥ If ▯ ▱ ▱ ▱ ▱ ▱ is 0.1, then what is the whole? _____ What is 0.01?

_____

⑦ Explain how you solved Problem 6. _____

_____

_____

⑧ Ramon said the value of the cube in Problem 2 was equivalent to the value of the cube in Problem 3. Do you agree or disagree? _____ Explain.

_____

_____

_____

Hundredths with Coins

①

Ⓠ Ⓓ Ⓝ Ⓟ

How many pennies? __41__

$\frac{41}{100}$ = 0.__41__

②

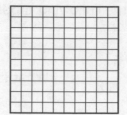

Ⓠ Ⓠ Ⓠ Ⓓ

How many pennies? _____

$\frac{\boxed{}}{100}$ = 0._____

③

SRB
150-151

Ⓓ Ⓓ Ⓓ Ⓝ Ⓟ

How many pennies? _____

$\frac{\boxed{}}{100}$ = 0._____

④

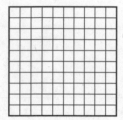

Ⓝ Ⓝ Ⓝ Ⓟ Ⓟ

How many pennies? _____

$\frac{\boxed{}}{\boxed{}}$ = 0._____

⑤

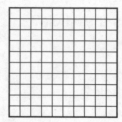

Ⓠ Ⓠ Ⓓ Ⓝ Ⓝ

How many pennies? _____

$\frac{\boxed{}}{\boxed{}}$ = 0._____

⑥

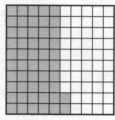

◯ ◯ ◯ ◯

How many pennies? _____

$\frac{\boxed{}}{\boxed{}}$ = 0._____

⑦

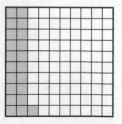

◯ ◯ ◯ ◯ ◯

How many pennies? _____

$\frac{\boxed{}}{\boxed{}}$ = 0._____

⑧

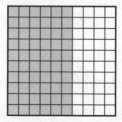

◯ ◯ ◯ ◯

How many pennies? _____

$\frac{\boxed{}}{\boxed{}}$ = 0._____

⑨ Create your own.

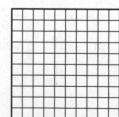

◯ ◯ ◯ ◯ ◯

$\frac{\boxed{}}{\boxed{}}$ = 0._____

Representing Fractions and Decimals

If the grid is the whole, then what part of each grid is shaded?

Write a fraction and a decimal below each grid.

①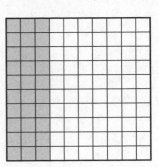

fraction: _____

decimal: _____

②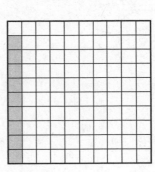

fraction: _____

decimal: _____

③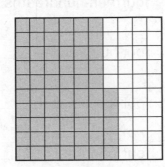

fraction: _____

decimal: _____

④ Color 0.8 of the grid.

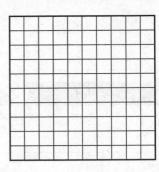

⑤ Color 0.04 of the grid.

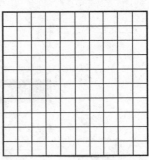

⑥ Color 0.53 of the grid.

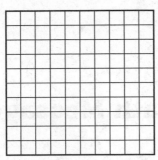

Practice

⑦ The numbers 81, 27, and 45 are all multiples of 1, _____, and _____.

⑧ List the first ten multiples of 6.

_____, _____, _____, _____, _____, _____, _____, _____,

_____, _____

Math Message

Write the following numbers with digits.

(1) fourteen-hundredths _____

(2) four-hundredths _____

(3) four-tenths _____

(4) four hundred _____

(5) forty-hundredths _____

(6) forty _____

- -

Math Message

Write the following numbers with digits.

(1) fourteen-hundredths _____

(2) four-hundredths _____

(3) four-tenths _____

(4) four hundred _____

(5) forty-hundredths _____

(6) forty _____

130

Place-Value Puzzles

Use the clues to write the digits in the boxes and find each number.

① • Write the result of 156 − 148 in the tenths place.

• Find half of 80. Divide by 8. Write the result in the ones place.

1s	0.1s	0.01s	0.001s
.			

• Add 41 to the digit in the tenths place. Divide by 7. Write the result in the hundredths place.

• In the thousandths place, write a digit larger than the digit in the tenths place.

② • Divide 100 by 20. Write the result in the hundredths place.

• Add 5 to the digit in the hundredths place and divide by 5. Write the result in the ones place.

10s	1s	0.1s	0.01s	0.001s
	.			

• Write a digit in the tenths place that is 4 more than the digit in the hundredths place.

• Write the first prime number after 6 in the thousandths place.

• Write the largest even digit not used yet in the tens place.

③ • Write the result of 6 * 9 divided by 18 in the ones place.

10s	1s	0.1s	0.01s	0.001s
	.			

• Double 8. Divide by 4. Write the result in the thousandths place.

• Add 3 to the digit in the thousandths place. Write the result in the tens place.

• Write the same digit in the tenths and hundredths places so that the sum of all the digits is 14.

④ Make up and solve a place-value puzzle of your own. Share it with a classmate.

Tenths and Hundredths

Family Note Your child continues to work with decimals. Encourage him or her to think about ways to write money amounts. This is called dollars-and-cents notation. For example, $0.07 (7 cents), $0.09 (9 cents), and so on.

Write the decimal numbers that represent the shaded part in each diagram.

Whole
grid

(1)

_____ hundredths

____ tenths ____ hundredths

(2)

_____ hundredths

____ tenths ____ hundredths

(3)

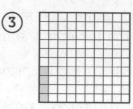

_____ hundredths

____ tenths ____ hundredths

SRB
149-150

Write the words as decimal numbers.

(4) twenty-three hundredths

_____

(5) eight and four-tenths

_____

(6) thirty and twenty-hundredths

_____

(7) five-hundredths

_____

Continue each pattern.

(8) 0.1, 0.2, 0.3, _____, _____, _____, _____, _____, _____

(9) 0.01, 0.02, 0.03, _____, _____, _____, _____, _____, _____

Practice

(10) Round 7,604 to the nearest thousand. _____

(11) Round 46,099 to the nearest thousand. _____

(12) Round 8,500,976 three ways: nearest thousand, hundred-thousand, and million.

_____ _____ _____

Representing Fractions and Decimals

Represent each fraction and decimal amount on the grid and on the number line.

Fraction: _____

Decimal: _____

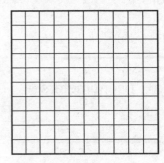

0   1

Fraction: _____

Decimal: _____

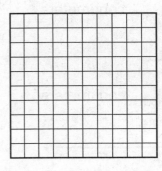

0   1

Fraction: _____

Decimal: _____

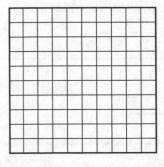

0   1

Measuring in Centimeters

Measure each line segment to the nearest centimeter. Record the measurement in centimeters and meters.

SRB
180

Example: _____

    a. About _____5_____ centimeters   **b.** About ___0.05___ meter

① _____

    a. About _____ centimeters   **b.** About _____ meter

② _____

    a. About _____ centimeters   **b.** About _____ meter

③ _____

    a. About _____ centimeters   **b.** About _____ meter

④ _____

    a. About _____ centimeters   **b.** About _____ meter

⑤ _____

    a. About _____ centimeters   **b.** About _____ meter

⑥ _____

    a. About _____ centimeters   **b.** About _____ meter

⑦ _____

    a. About _____ centimeters   **b.** About _____ meter

Practice with Decimals

Fill in the missing numbers.

①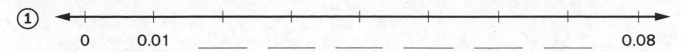

0   0.01   ____ ____ ____ ____ ____ ____   0.08

②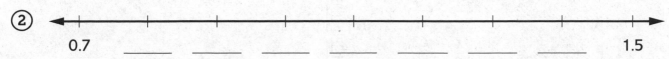

0.7   ____ ____ ____ ____ ____ ____   1.5

Follow these directions on the ruler below.

SRB
154,
182-183

③ Make a dot at 7 cm and label it with the letter *A*.

④ Make a dot at 90 mm and label it with the letter *B*.

⑤ Make a dot at 0.13 m and label it with the letter *C*.

⑥ Make a dot at 0.06 m and label it with the letter *D*.

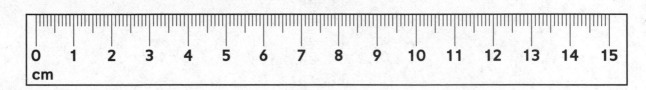

⑦ Write <, >, or =.

a. 1.2 ____ 0.12   **b.** 0.3 ____ 0.38   **c.** 0.80 ____ 0.08

⑧ Complete.

1 cm = 10 mm	1 m = 100 cm

cm	m
100	1
	5
1,000	
6,000	

cm	m
1	*0.01*
	0.03
	0.06
40	

Practice

⑨ 6,366 + 7,565 = _____

⑩ 3,238 + 29,784 = _____

⑪ 9,325 − 7,756 = _____

⑫ 14,805 − 2,927 = _____

135

Precipitation Gauge

22 — cm
21 —
20 —
19 —
18 —
17 —
16 —
15 —
14 —
13 —
12 —
11 —
10 —
9 —
8 —
7 —
6 —
5 —
4 —
3 —
2 —
1 —
0 —

Comparing Millimeters and Centimeters

cm: 0 1 2 3 4 5 6 7 8 9 10 11 12 13 14 15 16 17 18 19 20 21
mm: 0 10 20 30 40 50 60 70 80 90 100 110 120 130 140 150 160 170 180 190 200 210

This is a picture of a centimeter strip. It shows both centimeter (cm) and millimeter (mm) marks.

The top numbers (from 0–21) are centimeters. Underneath the centimeter numbers, there are larger numbers that tell the number of millimeters.

(1) Count the intervals from 0 to 1 cm to see that there are 10 millimeters for every 1 centimeter. Each millimeter is worth $\frac{1}{10}$, or 0.1, of a centimeter. So, 2.3 cm = 23 mm, and 37 mm = 3.7 cm.

(2) Find each of the points below and mark them on the centimeter strip with a colored pencil. Use the millimeter marks to help you convert the centimeters to millimeters. Remember to count intervals, not marks, to find the missing numbers.

a. 2 cm = _____ mm

b. 11 cm = _____ mm

c. 1.4 cm = _____ mm

d. _____ cm = 80 mm

e. _____ cm = 65 mm

f. _____ cm = 135 mm

Measuring in Millimeters

Measure each line segment to the nearest tenth of a centimeter.
Record the measurement in centimeters and millimeters.

Example: _____

   a. About ___12.5___ centimeters   **b.** About ___125___ millimeters

① _____

   a. About _____ centimeters   **b.** About _____ millimeters

② _____

   a. About _____ centimeters   **b.** About _____ millimeters

③ _____

   a. About _____ centimeters   **b.** About _____ millimeters

④ _____

   a. About _____ centimeters   **b.** About _____ millimeters

⑤ _____

   a. About _____ centimeters   **b.** About _____ millimeters

⑥ _____

   a. About _____ centimeters   **b.** About _____ millimeters

⑦ _____

   a. About _____ centimeters   **b.** About _____ millimeters

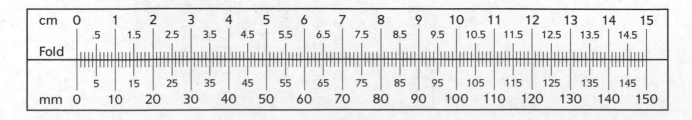

Measuring Centimeters and Millimeters

(1) Find 6 objects in your home to measure. Use the ruler from the bottom of the page to measure them, first in centimeters and then in millimeters. Record your objects and their measurements.

SRB
180, 182,-183

Example: _____*crayon*_____   ___*3.5*___ cm   ___*35*___ mm

Object   **Object**

_____ ____ cm ____ mm   _____ ____ cm ____ mm

_____ ____ cm ____ mm   _____ ____ cm ____ mm

_____ ____ cm ____ mm   _____ ____ cm ____ mm

Fill in the tables.

(2)

cm	mm
1	
15	
3.7	
49.6	
0.8	

(3)

cm	m
	1
180	
	23.6
	5.72
	0.65

Practice

(4) List the factors for 63. _____

(5) Write the factor pairs for 60.

_____ and _____   _____ and _____   _____ and _____

_____ and _____   _____ and _____   _____ and _____

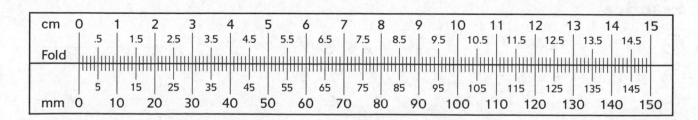

139

Comparing Decimals

Family Note Ask your child to read the decimal numerals aloud. Encourage your child to use the following method:

1. Read the whole-number part.
2. Say *and* for the decimal point.
3. Read the digits after the decimal point as though they form their own number.
4. Say *tenths* or *hundredths*, depending on the placement of the right-hand digit. Encourage your child to exaggerate the *-ths* sound. For example, 2.37 is read as "two and thirty-seven hundredths."

Write >, <, or =.

> means *is greater than*
>
> < means *is less than*

①   2.35 ____ 2.57   ②   1.08 ____ 1.8

③   0.64 ____ 0.46   ④   0.90 ____ 0.9

⑤   42.1 ____ 42.09   ⑥   7.09 ____ 7.54

⑦   0.4 ____ 0.40   ⑧   0.26 ____ 0.21

Example: The 4 in 0.47 stands for 4 __*tenths*__ or __*0.4*__.

⑨   The 9 in 4.59 stands for 9 _____ or _____.

⑩   The 3 in 3.62 stands for 3 _____ or _____.

Continue each number pattern.

⑪   6.56, 6.57, 6.58, _____, _____, _____

⑫   0.73, 0.83, 0.93, _____, _____, _____

Write the number that is 0.1 more.   Write the number that is 0.1 less.

⑬   4.3 _____   ⑭ 4.07 _____   ⑮ 8.2 _____   ⑯ 5.63 _____

Practice

⑰   43,589 + 12,641 = _____   ⑱   63,274 + 97,047 = _____

⑲   41,805 − 26,426 = _____   ⑳   82,004 − 11,534 = _____

Multidigit Multiplication

In Unit 4 your child will multiply multidigit numbers using **extended multiplication facts, partial-products multiplication,** and **lattice multiplication**. Throughout the unit, students use these methods to solve real-life multistep multiplication number stories.

The unit begins with extended multiplication facts. Knowing that $5 * 3 = 15$ helps students see that $50 * 3 = 150$; $500 * 3 = 1,500$; and so on. Working with extended facts gives students the ability to multiply larger numbers with ease.

Students also learn the partial-products multiplication method in which the value of each digit in one factor is multiplied by the value of each digit in the other factor. They partition a rectangle into smaller parts to help them understand how the method works. The example below shows how to use partial-products multiplication to find $456 * 4$.

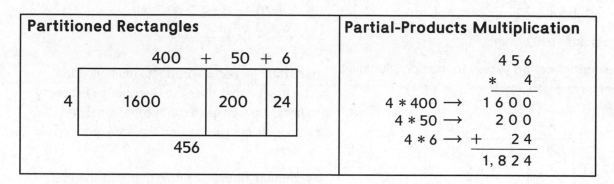

Partitioned Rectangles	Partial-Products Multiplication

To practice multiplying 2-digit numbers using partial-products multiplication, students play a game called *Multiplication Wrestling*.

Finally, students are introduced to the lattice multiplication method: The lattice method breaks down the numbers into place values, allowing students to work with smaller numbers while solving a multidigit multiplication problem. It is an efficient method, often taking no more time than other methods.

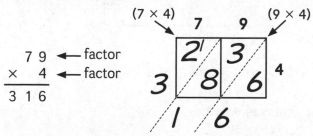

In this unit, students apply their understanding of multidigit multiplication to solve conversion problems involving liters and milliliters and grams and kilograms. They also find the area of rectilinear figures.

Please keep this Family Letter for reference as your child works through Unit 4.

Vocabulary

Important terms in Unit 4:

adjacent Next to, or adjoining.

decompose To "break apart" numbers into friendlier numbers.

Distributive Property A rule saying that if *a*, *b*, and *c* are real numbers, then: $a * (b + c) = (a * b) + (a * c)$.

extended multiplication facts Multiplication facts involving multiples of 10, 100, and so on. For example, $400 * 6 = 2,400$ and $20 * 30 = 600$ are extended multiplication facts.

gram (g) A unit of mass in the metric system. There are about 454 grams in 1 pound.

kilogram (kg) 1,000 grams.

lattice multiplication A way to multiply multidigit numbers. *For example:*

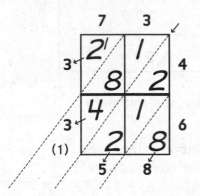

liter (L) A unit of capacity in the metric system. It is equivalent to a little more than one quart.

mass The measure of the amount of matter in an object.

milliliter (mL) $\frac{1}{1000}$ of a liter.

partial-products multiplication A way to multiply in which the value of each digit in one factor is multiplied by the value of each digit in the other factor. The final product is the sum of the partial products. *For example:*

```
                    7 3
             *      4 6
 40 * 70 →   2 8 0 0
 40 * 3  →     1 2 0
  6 * 70 →     4 2 0
  6 * 3  →  +    1 8
             3, 3 5 8
```

partition (in partial-products multiplication) A technique that uses the Distributive Property to break up a large rectangle into smaller rectangles in order to find the area more easily in parts.

rectilinear figure A single figure formed by combining multiple adjacent rectangles.

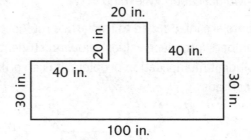

Do-Anytime Activities

To work with your child on concepts taught in this unit, try these activities:

1. Practice extended multiplication facts such as $50 * 40 =$ _____.

2. Collect three to five cans and bottles from the kitchen. Put them on the table and ask your child to order them, without looking at the labels, based on the amount of liquid each container can hold and/or their mass. Ask your child to estimate both. Check the results together by looking at the labels.

3. Pose a multiplication problem and ask your child to solve it using a method of his or her choice. Have your child explain to you or someone else at home what he or she did to complete the problem.

Building Skills through Games

In this unit your child will play the following game to develop his or her understanding of multiplication. For detailed instructions, see the *Student Reference Book*.

Multiplication Wrestling See *Student Reference Book*, page 267.
The game provides practice with multiplication of 2-digit numbers by 2-digit numbers.

As You Help Your Child with Homework

As your child brings assignments home, you may want to go over instructions together, clarifying them as necessary. The answers listed below will guide you through the Home Links for this unit.

Home Link 4-1
1. 560; 3,200; 630; 3,600

3. 450; 200; 63,000; 28,000

5. 9; 240; 700; 6,300

7. Answers vary.   9. 1,190

11. 13,303

Home Link 4-2
Number models are sample answers.

1. $(20 * 30) - (10 * 30) = 300$;
   330; Answers vary.

3. $30 * 50 = 1,500$; $30 * 40 = 1,200$; $1,500 - 1,200 = 300$; 496; Answers vary.

5. 1,410,000

Home Link 4-3
1. 140; Sample answer:

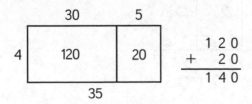

3. 441; Sample answer:

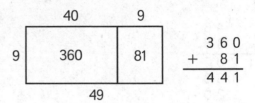

5. 2,956   7. 2,559

Home Link 4-4
1. 8,000; 15,000; 20,000; 25,000

3. 122,000 mL   5. 14,445   7. 62,341

Home Link 4-5
1. Sample answer: Four calculators fit in a layer. The box is 5 cm tall, so there are 5 layers of calculators. The box fits 4 calculators * 5, which is 20 calculators in all.

3. 108   5. 129

Home Link 4-6
1.
```
    4 8
  *   3
  -----
  1 2 0
  + 2 4
  -----
  1 4 4
```

3. 9 [100,000s] + 5 [1,000s] + 6 [100s] + 3 [1s]

5. 2 [1,000,000s] + 5 [100,000s] + 9 [10,000s] +
9 [1,000s] + 2 [1s]

Home Link 4-7

1. 25; 50,000; 75,000; 100

3. 237,000; 98,000; 485; 920,000

5. 63,000 grams   **7.** 396   **9.** 294

Home Link 4-8

1. $478   **3.** $55

5. 1, 3, 7, 21   **7.** 1, 2, 3, 4, 6, 9, 12, 18, 36

Home Link 4-9

1. 1,748

```
      4 6
  *   3 8
  1 2 0 0
    1 8 0
    3 2 0
  +     4 8
  1, 7 4 8
```

3. 65 * 22 = t; 1,430 trees

5. 185   **7.** 1,992

Home Link 4-10

1. 42; 420; 420; 4,200; 4,200; 42,000

3. 32; 320; 320; 3,200; 3,200; 32,000

5. 6; 6; 60; 9; 900; 9,000

7. 2,139   **9.** 32,632

Home Link 4-11

1. 18 * 27 = 486; 486 square units

3. Sample answer: 100 * 30 = 3,000;
20 * 20 = 400; 3,000 + 400 = 3,400;
3,400 square inches

5. 1, 2, 31, 62   **7.** 1, 5, 11, 55

Home Link 4-12

Sample number models:

1. (10 * 7) * 2 = 140; (5 * 7) * 2 = 70;
140 + 70 = 210 stickers;

(8 * 7) * 2 = x; (5 * 7) * 2 = y;
112 + 70 = s; 182 stickers

3. 1 and 50, 2 and 25, 5 and 10

5. 1 and 85, 5 and 17

Home Link 4-13

1. 536

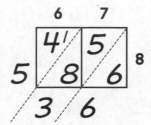

2. 5,852

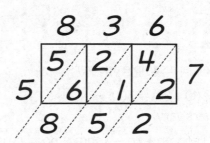

4. 2,552

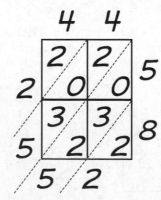

6. 616   **8.** 356

A 50-by-40 Array

Multiplication Puzzles

Solve the multiplication puzzles mentally. Fill in the blank boxes.

SRB 102

Examples:

*	300	2,000
2	600	4,000
3	900	6,000

*	80	50
4	320	200
8	640	400

①

*	70	400
8		
9		

②

*	5	7
80		
600		

③

*	9	4
50		
7,000		

④

*		600
7	3,500	
		2,400

⑤

*		8
30	270	
		5,600

⑥

*	400	
	3,600	
20		10,000

Make up and solve some puzzles of your own.

⑦

*		

⑧

*		

Practice

Solve using U.S. traditional addition or subtraction.

⑨ 321 + 869 = _____

⑩ 5,401 − 752 = _____

⑪ 4,568 + 8,735 = _____

⑫ 9,156 − 4,584 = _____

146

What Do You Eat?

Math Message

Answer the following questions as best you can.

(1) How many eggs did you eat in the last 7 days? _____ eggs

(2) How many cups of milk did you drink in the last 7 days? _____ cups

(3) How many cups of yogurt did you eat in the last 7 days? _____ cups

Use your answers to Problems 1–3 to complete these statements.

(4) I will eat about _____ eggs in 1 year.

(5) I will drink about _____ cups of milk in 1 year.

(6) I will eat about _____ cups of yogurt in 1 year.

(7) Based on your answers to Problems 4–6, do you eat the average amounts? _____

Explain why or why not on the back of this page.

What Do You Eat?

Math Message

Lesson 4-2

NAME   DATE   TIME

Answer the following questions as best you can.

(1) How many eggs did you eat in the last 7 days? _____ eggs

(2) How many cups of milk did you drink in the last 7 days? _____ cups

(3) How many cups of yogurt did you eat in the last 7 days? _____ cups

Use your answers to Problems 1–3 to complete these statements.

(4) I will eat about _____ eggs in 1 year.

(5) I will drink about _____ cups of milk in 1 year.

(6) I will eat about _____ cups of yogurt in 1 year.

(7) Based on your answers to Problems 4–6, do you eat the average amounts? _____

Explain why or why not on the back of this page.

147

Finding Missing Numbers and Digits

① Complete the number sentences.

SRB
103-107

• Fill in the circles using the numbers 3, 4, 6, or 7.

• Fill in the rectangles using the numbers 47, 62, 74, or 86.

• Some numbers will be used more than once.

a. ◯ × ▭ = 329

b. ◯ × ▭ = 258

c. ◯ × ▭ = 372

d. ◯ × ▭ = 248

e. ◯ × ▭ = 444

f. ◯ × ▭ = 296

For Problems 2–4, use each digit only once in each problem.

② For each problem, fill in the squares using the digits 4, 6, and 7.

a.
☐☐
× ☐
———
448

b.
☐☐
× ☐
———
322

c.
☐☐
× ☐
———
268

③ Use the digits 6, 7, 8, and 9 to make a problem with the largest product possible.

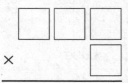

④ Use the digits 6, 7, 8, and 9 to make a problem with the smallest product possible.

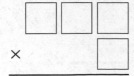

Finding Estimates and Evaluating Answers

Write an estimate and show your thinking. Solve using a calculator. Check to see that your answer is reasonable.

① Alice sleeps an average of 9 hours per night. A cat can sleep up to 20 hours per day. About how many more hours does a cat sleep in 1 month than Alice?

Estimate: _____

Answer: About _____ more hours per month

Is your answer reasonable? _____ How do you know? _____

_____

② Koalas sleep about 22 hours a day. Pandas sleep about 10 hours a day. About how many more hours does a typical koala sleep in 1 year than a typical panda?

Estimate: _____

Answer: About _____ more hours per year

Is your answer reasonable? _____ How do you know? _____

_____

③ There are 30 Major League Baseball (MLB) teams and 32 National Football League (NFL) teams. The expanded roster for MLB teams is 40 players and it is 53 for NFL teams. How many more players are in the NFL than in the MLB?

Estimate: _____

Answer: _____ more players

Is your answer reasonable? _____ How do you know? _____

_____

Practice

Round to the nearest thousand.

④ 45,493 _____

Round to the nearest ten-thousand.

⑤ 1,409,836 _____

149

Partitioning Rectangles

Solve the multiplication problems by partitioning a rectangle. Then add each part of the rectangle to get the product.

Example: 5 * 72 = _**360**_

① 4 * 35 = _____

```
       70        2
    ┌────────┬──────┐      350
  5 │  350   │  10  │    + 10
    └────────┴──────┘      360
          72
```

② 6 * 83 = _____

③ 9 * 49 = _____

Practice

Solve using U.S. traditional addition or subtraction.

④ 9,289 + 1,476 = _____

⑤ 6,503 − 3,547 = _____

⑥ 5,619 + 5,999 = _____

⑦ 5,005 − 2,446 = _____

150

Choosing Units to Measure Liquids

1 milliliter (mL)

—1 L

1 liter (L)

Units of Capacity

① Circle the unit you would use to measure each amount.

A large jug of milk   milliliters or liters

Water in a thimble   milliliters or liters

A glass of juice   milliliters or liters

Water in a water cooler   milliliters or liters

Water in a fish tank   milliliters or liters

Liquid in a paper cup   milliliters or liters

A tank of gas   milliliters or liters

A spoonful of oil   milliliters or liters

A large bottle of rruit juice   milliliters or liters

A can of soup   milliliters or liters

② Explain how you decided which unit to use for a can of soup.

_____

_____

Investigating Liters and Milliliters

Solve. Consider using a measurement scale or diagram.

Maria Elena's store sells soup in different-size containers.

Size of Container	Amount
Extra small	250 mL
Small	500 mL
Medium	1 L
Large	1.5 L
Extra large	2 L

(1) How much more soup does the extra large container hold than the extra small?

_____

(2) Lucas bought 5 containers of soup: 1 extra small, 3 small, and 1 medium.

How many milliliters of soup did he buy? _____ mL

(3) Kamu bought 3 large containers of soup. She needs 4,000 mL of soup.

a. Does she have enough? _____

b. If so, does she have extra soup? How much? _____

(4) Sally bought 1 extra small, 1 small, 1 large, and 2 extra large containers of soup.

How much soup did she buy? _____ mL

(5) Rocco wants to purchase 2.5 L of soup. However, the store is out of medium-size and extra large-size containers. Find two ways Rocco can buy 2.5 L of soup.

_____

_____

(6) Write and solve your own problem.

_____

_____

Converting Liquid Measures

Complete the table.

①

Liters (L)	Milliliters (mL)
8	
15	
20	
25	

② Mrs. Wong's students kept track of how much water they used to water the classroom plants. The first week they used 24 liters, and the second week they used 17 liters. How many more milliliters did they use the first week than the second?

Answer: _____ mL

③ My fish tank holds 64 liters of water. My neighbor's tank holds 58 liters of water. How many milliliters is that combined?

Answer: _____ mL

④ Mrs. Reyes filled her kiddie pool with 83 liters of water. Her children added 2,000 mL of water to the pool. How many liters of water are in the pool now?

Answer: _____ L

Practice

Solve using U.S. traditional addition or subtraction.

⑤ 4,638 + 9,807 = _____

⑥ 7,322 − 3,741 = _____

⑦ 55,812 + 6,529 = _____

⑧ 98,001 − 7,443 = _____

153

Walking Away with a Million Dollars

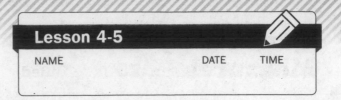

Imagine that you have done a heroic deed and will be given one million dollars as a reward. You get a box from your teacher to carry the money home.

Use what you learned in the Math Message and the following facts to solve the problems below.

A stack of bills 1 inch high has about 250 bills.

A ream of paper is 2 inches high.

The box holds 10 reams of paper.

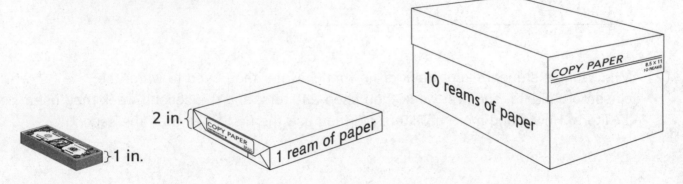

① Is it possible to fit the reward of one million dollars in $1 bills in your box? Show or explain how you know your answer makes sense.

Walking Away with a Million Dollars (continued)

② Is it possible to fit the reward of one million dollars in $100 bills in your box?
Show or explain how you know your answer makes sense.

Rubric for Walking Away with a Million Dollars

Goal: Check whether your answer makes sense.				
Meets Expectations	Student 1	Student 2	Student 3	
For at least one of the problems (either Problem 1 or Problem 2), the student used estimations, calculations, drawings, or a combination of these to clearly show how the problem was solved.	For Problem 1 **or** Problem 2:	For Problem 1 **or** Problem 2:	For Problem 1 **or** Problem 2:	
For at least one of the problems (either Problem 1 or Problem 2), it is easy to see how the numbers in the calculations or the diagrams are connected to the number of bills, sheets, reams, or dollar amounts.	For Problem 1 **or** Problem 2:	For Problem 1 **or** Problem 2:	For Problem 1 **or** Problem 2:	
Exceeds Expectations				
The student used two different strategies to check whether the answer makes sense.				

Using Multiplication

Ms. Patel wants to keep her classroom calculators in a box that is 25 centimeters long, 15 centimeters wide, and 5 centimeters tall. The calculators measure 12 centimeters long, 7 centimeters wide, and 1 centimeter tall. How many calculators can Ms. Patel fit in the box?

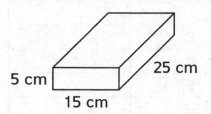

(1) Solve this problem. Show or explain how you solved the problem.

(2) Show or explain how you know your answer makes sense.

Practice

Sketch a rectangle or use partial products to solve.

(3) 27 * 4 = _____

(4) 48 * 9 = _____

(5) 43 * 3 = _____

(6) 81 * 6 = _____

157

Egyptian Multiplication

With a partner, carefully study the Egyptian multiplication algorithm below. Then solve a problem using this method.

Example: 13 × 28

Step 1: Write the first factor in the first column (13). Then write 1 in the first row below the factor. Double 1 and write 2 in the row below. Continue to double the number above until you get a number that is equal to or greater than the first factor. Cross out that number if it is greater than the first factor. 16 is crossed out.

1st factor: 13	2nd factor: ____
1	
2	
4	
8	
~~16~~	

Step 2: Write the second factor in the second column (28). Then write that number again in the box below. (It should be next to the 1 in the first column.) Double that number in each new line until the last number lines up with the last number of the first column. (224 lines up with 8.)

1st factor: 13	2nd factor: 28
1	28
2	56
4	112
8	224
~~16~~	

Step 3: Starting with the greatest number in column 1 (8), circle the numbers that add up to equal the first factor (13). 8 + 4 + 1 = 13

Cross out the row of numbers that you did not use to make the first factor (2 and 56).

1st factor: 13	2nd factor: 28
①	28
~~2~~	~~56~~
④	112
⑧	224
~~16~~	

Step 4: Add the numbers in the second column that are not crossed out. 28 + 112 + 224 = 364

Answer: 13 × 28 = 364

Check the answer by solving the problem using a method you already know.

1st factor: 13	2nd factor: 28
①	28
~~2~~	~~56~~
④	112
⑧	224
~~16~~	

Multiplication Match Directions

Work with a partner. You need Multiplication Match Cards, scissors, and tape.

SRB
103-104,
106

(1) Cut apart the cards on Multiplication Match Cards 1 and 2.

(2) Match each partitioned rectangle card with a partial-products card.

(3) When you have matched up all the cards, tape them in pairs on the front and back of this paper.

159

Multiplication Match Cards 1

	60	7
5	300	35

67

	800	80	7
2	1600	160	14

887

	10	8
3	30	24

18

	30	5
9	270	45

35

	20	7
5	100	35

27

	40	9
7	280	63

49

	30	8
7	210	56

38

	900	20	5
8	7200	160	40

925

	40	7
6	240	42

47

	200	6
4	800	24

206

	600	40
6	3600	240

640

	50	3
5	250	15

53

Copyright © McGraw-Hill Education. Permission is granted to reproduce for classroom use.

160

Multiplication Match Cards 2

```
   6 7          2 7          4 7        2 0 6
 *   5        *   5        *   6      *     4
 ─────        ─────        ─────      ───────
 3 0 0        1 0 0        2 4 0      8 0 0
 +   3 5      +   3 5      +   4 2    +   2 4
 ───────      ───────      ───────    ───────
 3 3 5        1 3 5        2 8 2      8 2 4
```

```
   4 9        8 8 7          1 8          3 8
 *   7      *     2        *   3        *   7
 ─────      ───────        ─────        ─────
 2 8 0      1 6 0 0          3 0        2 1 0
 +   6 3      1 6 0        +   2 4      +   5 6
 ───────    +     1 4      ───────      ───────
 3 4 3      ─────────        5 4        2 6 6
           1, 7 7 4
```

```
 6 4 0          5 3        9 2 5          3 5
 *     6      *   5      *     8        *   9
 ───────      ─────      ───────        ─────
 3 6 0 0      2 5 0      7 2 0 0        2 7 0
 +   2 4 0    +   1 5      1 6 0        +   4 5
 ─────────    ───────    +     4 0      ───────
 3, 8 4 0      2 6 5      ─────────      3 1 5
                         7, 4 0 0
```

161

Multiplying in Parts

In the example, a rectangle was drawn to represent the problem. Then partial-products multiplication was used to record the work in a simpler way. Use partial-products multiplication to solve Problems 1 and 2.

Example:

Partitioned Rectangle

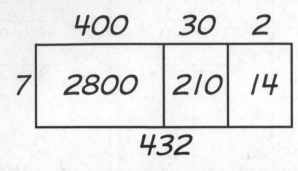

	400	30	2
7	2800	210	14

432

Partial-Products Multiplication

```
    4 3 2
  *     7
  2800
   210
+   14
  3,024
```

①
```
    4 8
  *   3
```

②
```
    6 5 3
  *     8
```

Practice

Write the numbers in expanded form.

③ 905,603 _____

④ 589,043 _____

⑤ 2,599,002 _____

⑥ 8,003,952 _____

Mammal Weights and Food Intake

Greta is doing research for a school project on marine mammals and the amount of food they eat, or their daily food intake. Help Greta fill in her chart by converting the kilogram amounts to grams.

Marine Mammal	Body Weight in Kilograms (kg)	Body Weight in Grams (g)	Daily Food Intake in Kilograms (kg)	Daily Food Intake in Grams (g)
Beluga Whale	1,000		20	
Bottlenose Dolphin	175		5	
California Sea Lion	275		7	
Harbour Seal	85		3	
Caribbean Manatee	400		80	
Northern Sea Lion	650		14	
Northern Elephant Seal	1,100		22	
Walrus	1,400		27	
Blue Whale	105,000		920	
Killer Whale	3,500		130	

Using a
Measurement Scale

(1) Fill in the blanks on the measurement scale.

kg 0 _____ 50 _____ 75 _____

g 0 _____ 25,000 _____ 100,000 _____

Complete the two-column tables.

(2)

Kilograms (kg)	Grams (g)
6	
14	
	27,000
101	

(3)

Kilograms (kg)	Grams (g)
237	
98	
	485,000
920	

(4) Find three items in your home that have the mass listed in grams or kilograms. Be sure to tell whether the mass is kilograms or grams.

Item	Mass in Kilograms (kg) or Grams (g)

(5) Among other foods, a giraffe in a zoo eats 4 kg of plant pellets and 5 kg of hay each day. How many grams of these foods does a giraffe eat in one week?

Answer: _____ grams

Practice

(6) $52 * 7 =$ _____

(7) $99 * 4 =$ _____

(8) $61 * 8 =$ _____

(9) $49 * 6 =$ _____

Edzo's Electronics Store

Edzo's Electronics has holiday sale items (including tax) as listed in the table below. Use the table to solve the multistep number stories. Write equations to show your work.

Product	Price	Product	Price
Alarm clock	$26	Headphones	$45
Blu-ray player	$69	Mini speakers	$11
Calculator	$83	Mobile phone	$179
Camera	$319	Printer	$149
Computer	$449	Television (40″)	$599
GPS unit	$119	Video game player	$199

① Jason is buying a new phone, an alarm clock, and a camera. If he gives the salesperson $550, what is his change?

Answer: $_____

② Dominique has $1,000 to buy a television, a video game player, and a Blu-ray player. Estimate to see if she has enough money. Then find the actual price she will pay.

Estimate: $_____

Does she have enough money? _____

Answer: $_____

③ Mrs. Murphy is buying each of her 3 daughters a GPS unit. She is also buying a computer for herself. What will she pay all together?

Answer: $_____

④ Using the information in the table, write a multistep number story on the back of this sheet and solve it.

Money Number Stories

Family Note Today your child solved multistep number stories involving multiplication, addition, and subtraction of money amounts. Have your child explain a plan for solving each of the following problems and then solve it.

Mr. Russo is buying equipment for his baseball team. Use the table to the right to answer questions about his purchases.

(1) Mr. Russo needs 9 helmets and 8 gloves. How much will they cost in all?

Answer: $_____

Item	Price
Wooden bat	$49
Metal bat	$74
Glove	$35
Helmet	$22

(2) Mr. Russo wants to buy 6 bats for his team. How much more would it cost for him to buy 6 metal bats than 6 wooden bats?

Answer: $_____

(3) Mr. Russo buys 5 wooden bats and gives the cashier $300. How much change does he get?

Answer: $_____

(4) If the cashier only has $10 and $1 bills, what are two ways he could make Mr. Russo's change?

Answer: _____

_____

Practice

List the factors for the following numbers:

(5) 21 _____ _____ _____

(6) 40 _____ _____ _____ _____ _____ _____

(7) 36 _____ _____ _____ _____

(8) 45 _____ _____ _____ _____

Multiplication Match Cards 1 (Advanced)

2 8
* 1 5
2 0 0
8 0
1 0 0
+   4 0
4 2 0

```
      2 8
   *  1 5
   2 0 0
     8 0
   1 0 0
 +   4 0
   4 2 0
```

```
      7 8
   *  3 9
   2 1 0 0
     2 4 0
     6 3 0
 +     7 2
   3,0 4 2
```

```
      3 7
   *  5 6
   1 5 0 0
     3 5 0
     1 8 0
 +     4 2
   2,0 7 2
```

```
      5 5
   *  7 5
   3 5 0 0
     3 5 0
     2 5 0
 +     2 5
   4,1 2 5
```

```
      3 8
   *  8 3
   2 4 0 0
     6 4 0
       9 0
 +     2 4
   3,1 5 4
```

```
      1 5
   *  1 5
   1 0 0
     5 0
     5 0
 +   2 5
   2 2 5
```

```
      4 4
   *  2 0
   8 0 0
 +   8 0
   8 8 0
```

```
      1 4
   *  9 7
   9 0 0
   3 6 0
     7 0
 +   2 8
   1,3 5 8
```

```
      7 5
   *  2 8
   1 4 0 0
     1 0 0
     5 6 0
 +     4 0
   2,1 0 0
```

167

Multiplication Match Cards 2 (Advanced)

	40	4
20	800	80

44

	20	8	
10	200	80	
5	100	40	15

28

	30	8	
80	2400	640	
3	90	24	83

38

	10	4	
90	900	360	
7	70	28	97

14

	70	8	
30	2100	240	
9	630	72	39

78

	50	5	
70	3500	350	
5	250	25	75

55

	30	7	
50	1500	350	
6	180	42	56

37

	70	5	
20	1400	100	
8	560	40	28

75

	10	5	
10	100	50	
5	50	25	15

15

Practicing Partial-Products Multiplication

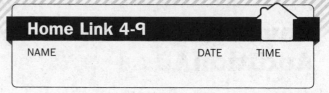

Solve using partial-products multiplication.

(1) 46 * 38 = _____

(2)
$$
\begin{array}{r}
6\ 5 \\
*\ 3\ 2 \\
\hline
\end{array}
$$

SRB
106-107

(3) Donnie and Raj went apple picking at an orchard that had 65 rows of trees. Each row had 22 trees in it. How many trees were in the orchard?

Number model with unknown: _____

Answer: _____ trees

(4) A new apartment building has 33 floors, with 24 apartments on each floor. How many apartments are in the building?

Number model with unknown: _____

Answer: _____ apartments

Practice

(5) 37 * 5 = _____

(6) 27 * 6 = _____

(7) 332 * 6 = _____

(8) 2,958 * 7 = _____

Reviewing Partial-Sums Addition

Example: 2,000 + 280 + 300 + 42 = ?

```
                              2, 0 0 0
                                 2 8 0
                                 3 0 0
                            +      4 2
Add the thousands:            2 0 0 0
Add the hundreds:               5 0 0   (200 + 300)
Add the tens:                   1 2 0   (80 + 40)
Add the ones:               +         2
Find the total:               2, 6 2 2
```

Solve each problem.

① 800 + 120 + 160 + 24 = _____

② 700 + 420 + 50 + 30 = _____

③ _____ = 600 + 180 + 40 + 12

④ _____ = 2,400 + 160 + 420 + 28

⑤ _____ = 1,500 + 90 + 240 + 24

⑥ 5,600 + 420 + 400 + 30 = _____

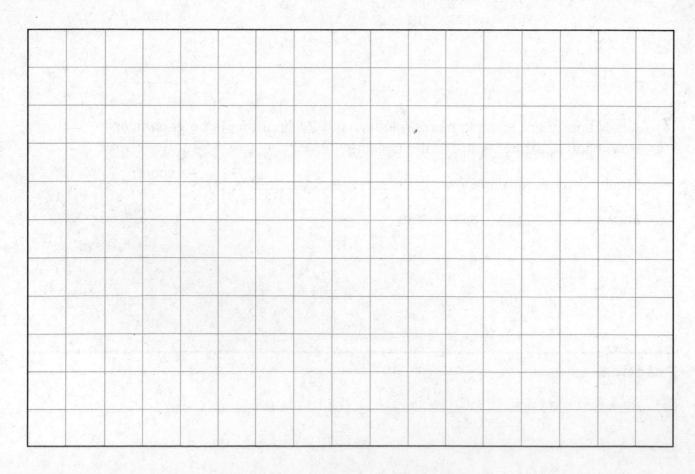

A Multiplication Wrestling Competition

① Twelve players entered a *Multiplication Wrestling* competition. The numbers they chose are shown in the following table. The score of each player is the product of the two numbers. For example, Aidan's score is 741 because 13 * 57 = 741. Which of the 12 players do you think has the highest score?

_____

Group A	Group B	Group C
Aidan: 13 * 57	Indira: 15 * 73	Miguel: 17 * 35
Colette: 13 * 75	Jelani: 15 * 37	Rex: 17 * 53
Emily: 31 * 75	Kuniko: 51 * 37	Sarah: 71 * 53
Gunnar: 31 * 57	Liza: 51 * 73	Tanisha: 71 * 35

Check your guess with the following procedure. *Do not do any arithmetic for Steps 2 and 3.*

② In each pair below, cross out the player with the lower score. Find that player's name in the table above and cross it out as well.

Aidan; Colette	Indira; Jelani	Miguel; Rex
Emily; Gunnar	Kuniko; Liza	Sarah; Tanisha

③ Two players are left in Group A. Cross out the one with the lower score.
Two players are left in Group B. Cross out the one with the lower score.
Two players are left in Group C. Cross out the one with the lower score.

Which 3 players are still left?

_____

④ Of the 3 players who are left, which player has the lowest score? _____
Cross out that player's name.

⑤ There are 2 players left. What are their scores? _____

_____

⑥ Who won the competition? _____

Finding Multiplication Wrestling Errors

Multiplication Wrestling Record Sheet

Round 1 Cards: _____5_____ _____9_____ _____6_____ _____3_____

Numbers formed: _____65_____ * _____93_____

Teams: (_____60_____ + _____5_____) * (_____90_____ + _____3_____)

Products: _____60_____ * _____9_____ = _____540_____

_____60_____ * _____3_____ = _____180_____

_____60_____ * _____5_____ = _____300_____

_____5_____ * _____90_____ = _____450_____

Total (add 4 products): _____1,470_____

Round 2 Cards: _____4_____ _____7_____ _____8_____ _____9_____

Numbers formed: _____94_____ * _____87_____

Teams: (_____90_____ + _____4_____) * (_____80_____ + _____7_____)

Products: _____90_____ * _____80_____ = _____7,200_____

_____90_____ * _____7_____ = _____630_____

_____80_____ * _____90_____ = _____7,200_____

_____80_____ * _____4_____ = _____320_____

_____15,350_____

Total (add 4 products):

Round 3 Cards: _____5_____ _____8_____ _____4_____ _____2_____

Numbers formed: _____54_____ * _____82_____

Teams: (_____50_____ + _____4_____) * (_____80_____ + _____2_____)

Products: _____50_____ * _____80_____ = _____4,000_____

_____50_____ * _____2_____ = _____100_____

_____4_____ * _____80_____ = _____320_____

_____4_____ * _____2_____ = _____8_____

_____7,308_____

Total (add 4 products):

Extended Multiplication Facts

Solve mentally.

SRB
56-57,
102

(1) 6 * 7 = _____

6 * 70 = _____

60 * 7 = _____

60 * 70 = _____

600 * 7 = _____

60 * 700 = _____

(2) 5 * 6 = _____

5 * 60 = _____

50 * 6 = _____

50 * 60 = _____

500 * 6 = _____

50 * 600 = _____

(3) 4 * 8 = _____

4 * 80 = _____

40 * 8 = _____

40 * 80 = _____

400 * 8 = _____

40 * 800 = _____

(4) 5 * _____ = 15

30 * _____ = 150

30 * _____ = 1,500

_____ * 50 = 150

_____ * 500 = 1,500

30 * _____ = 15,000

(5) 54 is _____ times as many as 9.

540 is _____ times as many as 90.

5,400 is _____ times as many as 90.

540 is 60 times as many as _____.

5,400 is 6 times as many as _____.

54,000 is 6 times as many as _____.

Practice

Solve using U.S. traditional addition or subtraction.

(6) 6,419 + 7,809 = _____

(7) 8,045 − 5,906 = _____

(8) 76,543 + 84,086 = _____

(9) 65,409 − 32,777 = _____

Finding the Area

Math Message

Find the area.

Perimeter = 752 inches

← 367 inches →

Area: _____ square inches

- -

Finding the Area

Math Message

Find the area.

Perimeter = 752 inches

← 367 inches →

Area: _____ square inches

Area and Perimeter of a Tennis Court

| Area of rectangle = $l * w$ | Perimeter of rectangle = $2 * (l + w)$ or $l + l + w + w$ |

When 2 people play tennis, it is called *singles*. When 4 people play, it is called *doubles*. In doubles, it is 2 players playing against 2 players. Below is a diagram of a tennis court. The net divides the court in half. The two alleys on the sides are used only in doubles.

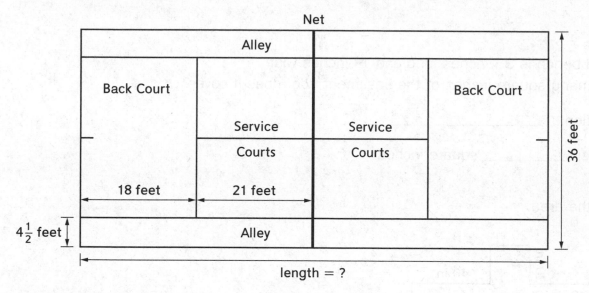

① What is the total length of a tennis court? _____ feet

② The court used in a game of doubles is 36 feet wide. Each alley is $4\frac{1}{2}$ feet wide.

What is the width of the court used in a game of singles? _____ feet

③ Find the perimeter and area of a singles court. Show your work on the back of this page.

Perimeter: _____ feet   Area: _____ square feet

④ Find the perimeter and area of a doubles court.

Perimeter: _____ feet   Area: _____ square feet

⑤ Do you think a player needs to cover more court in a game of singles or a game of doubles? Explain. _____

_____

_____

_____

Finding the Area

① Find the area.

Equation: _____

Answer: _____ square units

27

18

② A tool bench is 35 inches long and 19 inches wide.
How many square inches of the basement floor does it cover?

Equation: _____

Answer: _____ square inches

③ Find the area.

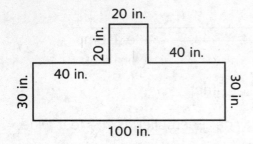

20 in.

20 in.

40 in.

40 in.

30 in.

30 in.

100 in.

Equations: _____

_____

Answer: _____ square inches

Practice

List all of the factors for the numbers below.

④ 48 _____

⑤ 62 _____

⑥ 63 _____

⑦ 55 _____

Solving Multistep Number Stories

Work with your partner to make sense of each problem and make a plan to solve it. Use the space below each problem to show your work and solution. Ask: *Does my answer make sense?*

(1) Jim's doctor said he should walk about 10,000 steps each day for his health, so Jim bought a pedometer to record his steps. About 2,120 of Jim's steps equal a mile. If Jim has walked about 4 miles today, how many steps has his pedometer recorded so far? How many more steps does Jim need to reach his goal?

(2) Robby's Lawn Service must work in Mrs. Gilroy's yard quickly because of a 3:00 P.M. storm prediction. Robby thinks it will take about 30 minutes for each task: trimming the bushes, mowing the lawn, weeding the flower beds, edging the lawn, raking, and cleaning up. They will begin working at 1 P.M. How long will it take 1 employee to do the work? How long for 2 employees working together? If you were Robby, how many employees would you send?

(3) Chandra has 2 flowerbeds to fill with tulip bulbs. One flowerbed measures 24 feet by 6 feet, and the other bed measures 15 feet by 8 feet. Each bulb needs about a square foot of space. If Chandra wants each bed to be half red tulips and half white tulips, how many of each color bulb should she buy?

Multistep Multiplication Number Stories

Write estimates and number models for each problem. Then solve.

SRB
26,
36-37

① Rosalie is collecting stickers for a scrapbook. She collected 8 stickers per day for 2 weeks and then collected 5 stickers per day for 2 weeks. How many stickers has Rosalie collected?

Estimate: _____

Number models with unknowns:

_____

_____

Answer: _____ stickers

② Rashaad's sister gives him 2 packs of baseball cards per month. Each pack has 11 cards. She gives him 3 extra packs for his birthday. How many cards does Rashaad get in a year?

Estimate: _____

Number models with unknowns:

_____

_____

Answer: _____ cards

Does your answer make sense? Explain. _____

_____

Practice

Name all the factor pairs.

③ 50 _____

④ 72 _____

⑤ 85 _____

⑥ 90 _____

178

Napier's Rods

Scottish mathematician John Napier (1550–1617) devised a multiplication method using rods made of bone, wood, or heavy paper. These rods were used to solve multiplication and division problems.

Example 1:

$4 * 67 = 268$

	6	7
1	0 / 6	0 / 7
2	1 / 2	1 / 4
3	1 / 8	2 / 1
4	2 / 4	2 / 8
5	3 / 0	3 / 5
6	3 / 6	4 / 2
7	4 / 2	4 / 9
8	4 / 8	5 / 6
9	5 / 4	6 / 3

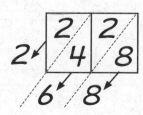

Example 2:

$8 * 5{,}239 = 41{,}912$

	5	2	3	9
1	0 / 5	0 / 2	0 / 3	0 / 9
2	1 / 0	0 / 4	0 / 6	1 / 8
3	1 / 5	0 / 6	0 / 9	2 / 7
4	2 / 0	0 / 8	1 / 2	3 / 6
5	2 / 5	1 / 0	1 / 5	4 / 5
6	3 / 0	1 / 2	1 / 8	5 / 4
7	3 / 5	1 / 4	2 / 1	6 / 3
8	4 / 0	1 / 6	2 / 4	7 / 2
9	4 / 5	1 / 8	2 / 7	8 / 1

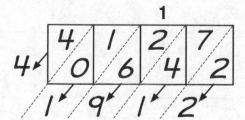

Cut out the rods on *Math Masters*, page 180. Use the rods and the board on *Math Masters*, page 181 to solve the following problems and some of your own. Use another method to check your answers.

(1) $5 * 79 = $ _____

(2) $7 * 92 = $ _____

(3) _____ $= 6 * 236$

(4) _____ $= 9 * 5{,}841$

Try This

(5) Show a friend how you would use Napier's Rods to solve $3 * 407$ or $9 * 5{,}038$.

179

Napier's Rods (continued)

1	2	3	4	5	6	7	8	9
0/1	0/2	0/3	0/4	0/5	0/6	0/7	0/8	0/9
0/2	0/4	0/6	0/8	1/0	1/2	1/4	1/6	1/8
0/3	0/6	0/9	1/2	1/5	1/8	2/1	2/4	2/7
0/4	0/8	1/2	1/6	2/0	2/4	2/8	3/2	3/6
0/5	1/0	1/5	2/0	2/5	3/0	3/5	4/0	4/5
0/6	1/2	1/8	2/4	3/0	3/6	4/2	4/8	5/4
0/7	1/4	2/1	2/8	3/5	4/2	4/9	5/6	6/3
0/8	1/6	2/4	3/2	4/0	4/8	5/6	6/4	7/2
0/9	1/8	2/7	3/6	4/5	5/4	6/3	7/2	8/1

Napier's Rods (continued)

1								
2								
3								
4								
5								
6								
7								
8								
9								

Lattice Multiplication

Use the lattice method to find the products.

Example   5 * 46 = _230_

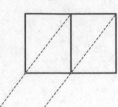

(1)   8 * 67 = _____

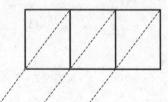

(2)   7 * 836 = _____

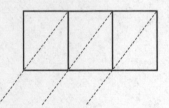

(3)   6 * 531 = _____

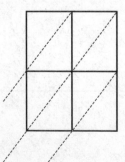

(4)   44 * 58 = _____

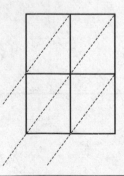

(5)   84 * 78 = _____

Practice

(6)   77 * 8 = _____

(7)   49 * 2 = _____

(8)   89 * 4 = _____

(9)   183 * 5 = _____

Fraction and Mixed-Number Computation; Measurement

In Unit 3 students learned how to compare and order fractions and decimals. In Unit 5 they deepen their understanding by learning how a fraction such as $\frac{3}{4}$ can be broken into smaller parts, such as $\frac{1}{4} + \frac{1}{4} + \frac{1}{4}$. Based on this understanding, students are able to see how adding and subtracting fractions with like denominators is simply putting together or taking away some number of same-size parts. For example, $\frac{3}{4} - \frac{1}{4}$ can be thought of as taking away 1 of the 3 parts, or fourths, that make up the fraction $\frac{3}{4}$.

In this unit students extend this idea to adding and subtracting mixed numbers, such as $1\frac{1}{4} + 2\frac{2}{4}$. They use different fraction representations and tools, including fraction circles, number lines, and drawings, to build a concrete understanding of the meaning of fractions, as opposed to just learning rules and procedures.

Line Plots

Line plots are used to organize and display data. As you can see from the diagram, a line plot can be thought of as a rough sketch of a bar graph.

From this line plot we can learn that the tallest books are 9 inches tall and that there are 3 of them, that no books are $8\frac{3}{4}$ inches tall, and so on.

Students also create line plots with data they collect in fractional units and then use information in the plots to solve problems involving adding and subtracting fractions.

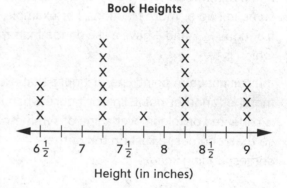

Book Heights

Height (in inches)

Angles: Unit Iteration and Rotations

Students begin their work with angle measurement by exploring the attribute of angle size. They begin measuring angles using a nonstandard unit—a wedge—as a way to see how measuring an angle is the same as measuring any other attribute. Iterating (or repeating) unit angles fills the spread between an angle's rays, just as iterating unit lengths fills a given length. Students discuss the need for a standard unit of measure, and they are introduced to the degree. An angle that measures 1 degree is a very small angle, which, when iterated 360 times, forms a circle.

Symmetry

Students complete symmetric figures that are partially given and create their own symmetric figures.

Multistep Multiplication Number Stories

Students continue solving multistep number stories, with a focus on multidigit multiplication strategies. They use number models that include a letter for the unknown, and they consider the reasonableness of their answers.

Please keep this Family Letter for reference as your child works through Unit 5.

Vocabulary

Important terms in Unit 5:

arc A part of a circle centered on the vertex of an angle. An arc is sometimes used to indicate where to measure the angle.

decompose To break apart a number or shape into smaller numbers or shapes.

degree A unit of measure for angles based on dividing a circle into 360 equal arcs.

full-turn A 360° rotation.

half-turn A 180° rotation.

like denominator A denominator that is the same in two or more fractions. For example, the fractions $\frac{3}{8}$, $\frac{5}{8}$, and $\frac{6}{8}$ have a like denominator, which is 8.

mirror image A point, line, or figure that exactly matches another point, line, or figure when it is reflected or folded over a line of symmetry so that it comes to rest on top of the corresponding image.

Sometimes the line of reflection is called a mirror, or mirror line.

mixed number A number that is written using both a whole number and a fraction.

quarter-turn A 90° rotation.

reflex angle An angle measure that is between 180° and 360°.

rotation A change in the direction an object faces; a turn.

straight angle An angle that measures 180°.

three-quarter turn A 270° rotation.

unit fraction A fraction whose numerator is 1. For example: $\frac{1}{2}$, $\frac{1}{3}$, and $\frac{1}{12}$ are unit fractions.

whole The entire object, collection of objects, or quantity being considered in a problem situation; 100%.

Do-Anytime Activities

To work with your child on concepts taught in this unit, try these activities:

1. Have your child help you measure when you are cooking or baking, using fractional measurements like $2\frac{1}{2}$ cups of flour or $\frac{1}{4}$ teaspoon of salt. Ask your child how you would double the measurements to make two batches instead of one. See whether he or she can show you one or two ways to do this.

2. Work with your child to create a line plot showing the number of hours family members spend sleeping or engaged in some other routine activity. Ask questions about the line plot; for example: "How many people in the family sleep for $8\frac{1}{2}$ hours?"

3. At home or when you are out together, encourage your child to point out items he or she believes are symmetric. Ask how many lines of symmetry there are in each of these objects.

4. Have your child point out angles in your home. Ask whether the angles are obtuse, acute, or right angles.

Building Skills through Games

In this unit, your child will play the following games as a way to increase his or her understanding of adding and subtracting fractions and mixed numbers, as well as angles, symmetry, and multistep multiplication number stories. For detailed instructions, see the *Student Reference Book*.

Decimal Top-It See *Student Reference Book*, page 253. This game provides practice comparing, ordering, reading, and identifying the values of digits in decimal numbers.

Fishing for Fractions (**Addition/Subtraction**) See *Student Reference Book*, page 260. In this game students practice adding together two like fractions or subtracting one like fraction from another.

Fraction/Decimal Concentration See *Student Reference Book*, page 262. This game helps students recognize when fractions and decimals are equivalent.

Fraction Match See *Student Reference Book*, page 263. This game develops skill in naming equivalent fractions.

Fraction Top-It See *Student Reference Book*, page 265. This game develops skill in comparing fractions.

As You Help Your Child with Homework

As your child brings assignments home, it may be helpful to review the instructions together, clarifying them as necessary. The answers listed below will guide you through some of the Home Links in Unit 5.

Home Link 5-1

1. Sample answers:
$\frac{1}{5} + \frac{1}{5} + \frac{1}{5} + \frac{1}{5} + \frac{1}{5} + \frac{1}{5} + \frac{1}{5} + \frac{1}{5} + \frac{1}{5} + \frac{1}{5} + \frac{1}{5}$;
$\frac{5}{5} + \frac{5}{5} + \frac{1}{5}$;
$\frac{5}{5} + \frac{6}{5}$;
$1 + 1 + \frac{1}{5}$; $\frac{2}{5} + \frac{3}{5} + \frac{6}{5}$

3. **a.** Sample answer: $\frac{4}{12} + \frac{4}{12} = \frac{8}{12}$;

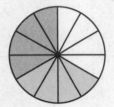

b. Sample answer: $\frac{8}{12} = \frac{2}{12} + \frac{2}{12} + \frac{2}{12} + \frac{2}{12}$;

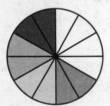

5. 3,227   7. 1,950

Home Link 5-2

1. **a.**   **b.**   **c.**

3. **a.**

b.

c.

d.

5. 4

Home Link 5-3

1. **a.** | Whole |
| new trees |

b. Sample answer: $\frac{1}{10} + \frac{3}{10} + \frac{2}{10} = t$

c. Sample answer:

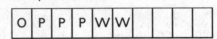

d. $\frac{6}{10}$ oak, willow, or pine

3. $\frac{3}{5}$ 5. $\frac{10}{6}$, or $1\frac{4}{6}$

7. 0.4   9. 0.6

Home Link 5-4

1. **a.** | Whole |
| ball of yarn |

b. Sample answer: $6\frac{2}{3} + 2\frac{2}{3} = s$

c. Sample answer: I decomposed the mixed numbers. Then I combined the wholes and the fractions. $6 + \frac{2}{3} + 2 + \frac{2}{3} = 8 + \frac{4}{3} = 9\frac{1}{3}$

d. $9\frac{1}{3}$, or $\frac{28}{3}$, balls

3. $8\frac{3}{6}$, or $\frac{51}{6}$

5. $6\frac{2}{4}$, or $\frac{26}{4}$ 7. 5,022   9. 1,092

Home Link 5-5

1. 20 hundredths + 15 hundredths = 35 hundredths

3. $\frac{10}{100} + \frac{50}{100} = \frac{60}{100}$, or $\frac{1}{10} + \frac{5}{10} = \frac{6}{10}$

5. $1 + \frac{30}{100} + 5 + \frac{64}{100} = 6\frac{94}{100}$

7. $\frac{150}{100} + \frac{78}{100} = \frac{228}{100}$, or $2\frac{28}{100}$

9. $\frac{2}{4}, \frac{3}{6}, \frac{4}{8}$

11. $\frac{2}{8}, \frac{3}{12}, \frac{4}{16}$

Home Link 5-6

1. Sample answer: Bill and Carl didn't get $\frac{1}{5}$. They each got $\frac{1}{8}$. I know this because the two triangles are in $\frac{1}{4}$ of the whole land, so each is half of a fourth, or $\frac{1}{8}$.

3. 15,732

5. 10,591

Home Link 5-7

1. **a.**

Whole
Elijah's allowance

b. Sample answer: $\frac{4}{5} - \frac{3}{5} = a$

c. Sample answer:

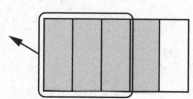

d. $\frac{1}{5}$ of his allowance

3. $\frac{1}{2}$

5. $\frac{4}{5}$

7. 2,243

9. 17,437

Home Link 5-8

1. **a.**

Whole
1 cup

b. Sample answer: $3\frac{1}{3} - 1\frac{2}{3} = c$

c. Sample answer: I started with what we had and counted up to what we needed. $1\frac{2}{3} + \frac{1}{3} = 2$. $2 + 1\frac{1}{3} = 3\frac{1}{3}$. Then I added: $\frac{1}{3} + 1\frac{1}{3} = 1\frac{2}{3}$.

d. $1\frac{2}{3}$, or $\frac{5}{3}$, cups

3. 1

5. $2\frac{3}{5}$, or $\frac{13}{5}$

7. 540

9. 8,084

Home Link 5-9

1.

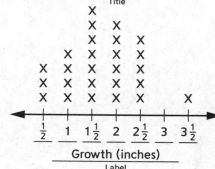

3. $3\frac{1}{2} - \frac{1}{2} = d$; 3 inches

5. $\frac{3}{4}, \frac{6}{8}, \frac{9}{12}$

7. $\frac{1}{2}, \frac{5}{10}, \frac{4}{8}$

Home Link 5-10

1–6. E, F, and G are sample answers

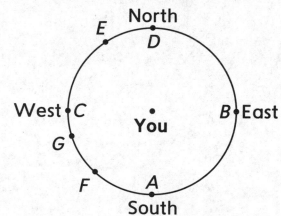

7. 4,250

9. 2,388

Home Link 5-11

1. angle *A*   **3.** angle *E*   **5.** angle *A* or *B*

7. 146,388   **9.** 12,961

Home Link 5-12

1. **a.** Triangle **b.** 2 sides **c.** 2 angles **d.** No.

3. 9   **5.** 6   **7.** 9

Home Link 5-13

1. Sample answer: 3 * (8 * 2 * 42) = *p*;
Sample answer: 3 * 800 = $2,400;
$2,016, or 2,016 dollars

3. 9   **5.** 6   **7.** 36

Decomposing Fractions Greater Than 1

① Which equations show $\frac{13}{8}$ decomposed? Write *True* or *False*.

Equation	True or False
$\frac{1}{8} + \frac{1}{8} + \frac{3}{8} + \frac{7}{8} = \frac{13}{8}$	
$\frac{13}{8} = \frac{4}{8} + \frac{4}{8} + \frac{4}{8} + \frac{1}{8}$	
$\frac{2}{8} + \frac{3}{8} + \frac{2}{8} + \frac{3}{8} + \frac{2}{8} = \frac{13}{8}$	
$\frac{13}{8} = \frac{6}{8} + \frac{6}{8} + \frac{2}{8}$	
$\frac{13}{8} = \frac{7}{8} + \frac{5}{8} + \frac{1}{8}$	
$\frac{8}{8} + \frac{3}{8} + \frac{1}{8} + \frac{1}{8} = \frac{13}{8}$	
$\frac{1}{8} + \frac{1}{8} + \frac{1}{8} + \frac{1}{8} + \frac{1}{8} + \frac{1}{8} + \frac{1}{8} + \frac{1}{8} + \frac{1}{8} + \frac{1}{8} + \frac{1}{8} + \frac{1}{8} = \frac{13}{8}$	

② Decompose the fraction $\frac{17}{12}$ into a sum of fractions with the same denominator using as many different equations as you can.

_____

_____

_____

_____

_____

_____

_____

_____

_____

_____

Exploring Tangrams

(1) Cut out the tangram pieces at the top of *Math Masters*, page TA41, and use all 7 pieces to create the large square at the bottom of page TA41. Trace the pieces to show your solution.

(2) If the large square is the whole, find the value of each of the tangram pieces.

Small Square	Large Triangle	Medium Triangle	Small Triangle	Parallelogram

(3) Describe the strategy you used to find the value of the small triangle.

_____

_____

_____

_____

(4) Describe how you can prove that you found the correct value of the small triangle.

_____

_____

_____

_____

Try This

(5) Use several tangram pieces to create a polygon for which the small square is worth $\frac{2}{9}$. Trace the polygon on the back of this page. Give the value of each tangram piece in the polygon.

Decomposing Fractions

Family Note In class today your child learned to decompose fractions into smaller parts. For example, $\frac{5}{6}$ can be decomposed into $\frac{1}{6} + \frac{1}{6} + \frac{1}{6} + \frac{1}{6} + \frac{1}{6}, \frac{2}{6} + \frac{3}{6}, \frac{1}{6} + \frac{4}{6}$, and so on.

Complete the name-collection boxes using equations.

SRB 125-127

(1) $\frac{11}{5}$

(2) $1\frac{3}{8}$

(3) Decompose $\frac{8}{12}$ in more than one way into a sum of fractions with the same denominator.

Record each decomposition with an equation and justify it by shading the circle.

a. Equation: _____

b. Equation: _____

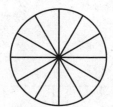   

Practice

(4) $9 * 785 =$ _____

(5) $461 * 7 =$ _____

(6) $644 * 4 =$ _____

(7) _____ $= 39 * 50$

Fractions of Rectangles

Use red, blue, and green crayons to color the squares at the bottom of the page. Cut out the squares. If your teacher has colored tiles, use those instead.

Use your colored squares to build the following rectangles in at least two *different* ways. Record your work.

① $\frac{1}{2}$ red and $\frac{1}{2}$ blue

② $\frac{1}{3}$ red, $\frac{1}{3}$ blue, $\frac{1}{3}$ green

③ $\frac{1}{4}$ red, $\frac{1}{2}$ blue, $\frac{1}{4}$ green

④ Make up a problem of your own.

_____ red, _____ blue, _____ green

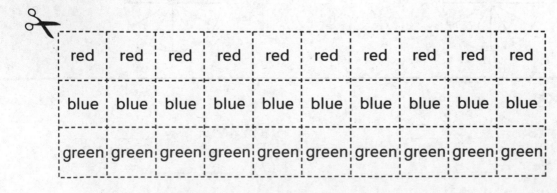

A Whole Candy Bar

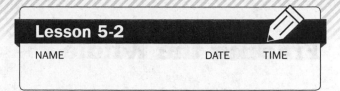
Two friends cut a large candy bar into equal pieces. Harriet ate $\frac{1}{4}$ of the pieces. Nisha ate $\frac{1}{2}$ of the remaining pieces. Six pieces were left over.

① How many pieces was the candy bar originally divided into? _____ pieces

② Explain how you got your answer. Include a drawing and equations as part of your explanation.

_____

_____

_____

_____

_____

✂ -

A Whole Candy Bar

Lesson 5-2

NAME   DATE   TIME

Two friends cut a large candy bar into equal pieces. Harriet ate $\frac{1}{4}$ of the pieces. Nisha ate $\frac{1}{2}$ of the remaining pieces. Six pieces were left over.

① How many pieces was the candy bar originally divided into? _____ pieces

② Explain how you got your answer. Include a drawing and equations as part of your explanation.

_____

_____

_____

_____

Finding the Whole

Use your Geometry Template to draw answers to these problems.

① If ⬦ is $\frac{1}{2}$, then what is the whole?

② If ⬥ is $\frac{1}{4}$, then what is the whole?

③ If ▭ is $\frac{2}{3}$, then what is the whole?

④ If ⬡ is $\frac{2}{5}$, then what is the whole?

Create problems of your own.

⑤

⑥

What Is the Whole?

For Problems 1-3, use your Geometry Template or sketch the shapes.

(1) Suppose [ ] is $\frac{1}{4}$. Draw each of the following:

Example: $\frac{3}{4}$ **a.** 1   **b.** $1\frac{1}{2}$ **c.** 2

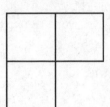

(2) Suppose ◇ is $\frac{2}{3}$. Draw each of the following:

a. $\frac{1}{3}$ **b.** 1   **c.** $\frac{4}{3}$ **d.** 2

(3) Suppose △ is $\frac{1}{3}$. Draw each of the following:

a. $\frac{3}{3}$ **b.** 2

c. $\frac{5}{3}$ **d.** $1\frac{1}{3}$

Practice

(4) $\frac{4}{5} = \frac{8}{\boxed{}}$ **(5)** $\frac{3}{\boxed{}} = \frac{9}{12}$ **(6)** $\frac{9}{10} = \frac{\boxed{}}{100}$

Investigating Egyptian Fractions

Ancient Egyptians only used fractions with 1 in the numerator. These are called **unit fractions.** Egyptians wrote non-unit fractions, such as $\frac{3}{4}$ and $\frac{4}{9}$, as sums of unit fractions. They did not use the same unit fraction more than once in a sum.

Examples:

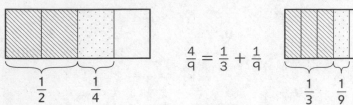

$$\frac{3}{4} = \frac{1}{2} + \frac{1}{4}$$

$$\frac{4}{9} = \frac{1}{3} + \frac{1}{9}$$

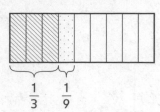

Use drawings and what you know about equivalent fractions to help you find the Egyptian form of each fraction.

(1) $\frac{3}{8} =$ _____

(2) $\frac{5}{12} =$ _____

(3) $\frac{7}{10} =$ _____

(4) $\frac{5}{6} =$ _____

(5) $\frac{3}{5} =$ _____

(6) $\frac{4}{7} =$ _____

Adding Fractions

Solve the number stories. Use a different strategy for each one.

SRB
47, 160–161

(1) The park department wants to have new trees planted. They agreed that $\frac{1}{10}$ of the trees will be oak, $\frac{3}{10}$ will be pine, and $\frac{2}{10}$ will be willow. They are undecided about the rest. What fraction of the trees will be oak, willow, or pine?

   a. Fill in the whole box.

Whole

   b. Number model with unknown:

   _____

   c. One way to solve a fraction addition problem:

   d. Answer (with unit): _____

(2) The Patels have a DVD collection. Three-eighths of the DVDs are animated. Two-eighths of them are mysteries. One-eighth are comedies. The rest are about travel. What fraction of the DVDs are *not* about travel?

   a. Fill in the whole box.

Whole

   b. Number model with unknown:

   _____

   c. A different way to solve a fraction addition problem:

   _____

   d. Answer (with unit): _____

Add.

(3) $\frac{2}{5} + \frac{1}{5} =$ _____

(4) $\frac{1}{2} + \frac{3}{2} =$ _____

(5) $\frac{5}{6} + \frac{5}{6} =$ _____

(6) $\frac{1}{3} + \frac{2}{3} + \frac{1}{3} =$ _____

Practice

Represent the fractions as decimals.

(7) $\frac{4}{10} =$ _____

(8) $\frac{40}{100} =$ _____

(9) $\frac{6}{10} =$ _____

(10) $\frac{6}{100} =$ _____

Adding Mixed Numbers with Unlike Denominators

Use the information in the table about state park trails to answer the questions below.

Trail	Miles	Type	Trail	Miles	Type
Cliff	$\frac{3}{4}$	rugged	Sky	$1\frac{1}{2}$	moderate
Pine	$\frac{3}{4}$	easy	Bluff	$1\frac{3}{4}$	rugged
Ice Age	$1\frac{1}{4}$	easy	Kettle	2	moderate
Oak	$1\frac{1}{2}$	easy	Badger	$3\frac{1}{2}$	moderate

① On his first day in the park, Luis wants to hike a total of 4 miles on easy or moderate trails. What combination of trails could he hike?

Number model: _____

Trail names: _____

② On Day 2, Luis decided to hike 5 miles on easy or moderate trails. Name two combinations of trails he could take.

Number models: _____

Trail names: _____

③ On Day 3, Luis would like to hike a total of 6 miles and wants to include at least one rugged trail. Give two combinations of trails he could hike.

Number models: _____

Trail names: _____

④ On his last day, Luis felt he was ready to do 2 different rugged trails and wanted to hike 8 miles. List the trails for two ways he could do this.

_____

_____

Mixed-Number Addition

Solve the number stories. Use a different strategy for each one.

① The art class had a box filled with balls of yarn. The students used $6\frac{2}{3}$ balls for a project. There are now $2\frac{2}{3}$ balls left in the box. How many balls of yarn did the art class start with?

    a. Fill in the whole box.   **b.** Number model with unknown:

    | Whole |
               _____

    c. One way to solve a mixed-number addition problem:

    d. Answer (with unit): _____

② Mrs. Meyers is growing vines along the sides of her house. On the west side the vines are $2\frac{4}{10}$ meters tall. On the east side the vines are $5\frac{8}{10}$ meters taller than the ones on the west side. How tall are the vines on the east side?

    a. Fill in the whole box.   **b.** Number model with unknown:

    | Whole |
               _____

    c. A different way to solve a mixed-number addition problem:

    d. Answer (with unit): _____

Add. Show your work.

③ $5\frac{2}{6} + 3\frac{1}{6} =$ _____

④ $1\frac{5}{8} + 2\frac{3}{8} =$ _____

⑤ $3\frac{3}{4} + 2\frac{3}{4} =$ _____

⑥ $3\frac{2}{5} + 1\frac{4}{5} + 2\frac{3}{5} =$ _____

Practice

⑦ $837 * 6 =$ _____

⑧ _____ $= 468 * 5$

⑨ _____ $= 364 * 3$

⑩ $56 * 70 =$ _____

Using Coins to Add Fractions

In the United States coins are all worth a fraction of a dollar. For example, a penny is worth $\frac{1}{100}$ of a dollar. The values of some coins can be expressed as a fraction of a dollar in different ways. A quarter can be expressed as $\frac{1}{4}$ of a dollar and as $\frac{25}{100}$ of a dollar.

When a coin's value is written as a *unit fraction* (a fraction with a numerator of 1), the denominator tells you how many of those coins are needed to make a dollar. When a coin's value is written as a fraction with a denominator of 100, the numerator tells you how many pennies are needed to equal the value of the coin.

Find the missing numbers to show the value of each coin as a fraction of a dollar.

Coin	Unit fraction	Denominator of 100
Penny:	$\frac{1}{\Box}$	$\frac{\Box}{100}$
Nickel:	$\frac{1}{\Box}$	$\frac{\Box}{100}$
Dime:	$\frac{1}{\Box}$	$\frac{\Box}{100}$
Quarter:	$\frac{1}{\Box}$	$\frac{\Box}{100}$
Half Dollar:	$\frac{1}{\Box}$	$\frac{\Box}{100}$

For Problems 1–3 on the next page, rewrite the equation to show the value of each coin as a unit fraction. Then rename the fractions as hundredths and add to find the total value of the coins.

Example:

$(P) + (Q) + (N) + (N) + (HD) = ?$

Unit fractions: $\frac{1}{100} + \frac{1}{4} + \frac{1}{20} + \frac{1}{20} + \frac{1}{2} = ?$

Hundredths: $\frac{1}{100} + \frac{25}{100} + \frac{5}{100} + \frac{5}{100} + \frac{50}{100} = \frac{86}{100}$, or 86 cents

Using Coins to Add Fractions (continued)

SRB
166-167

(1) (N) + (D) + (P) + (D) + (Q) = ?

Unit fractions: _____

Hundredths: _____

(2) (Q) + (Q) + (D) + (D) + (P) = ?

Unit fractions: _____

Hundredths: _____

(3) (P) + (N) + (D) + (Q) + (HD) = ?

Unit fractions: _____

Hundredths: _____

For Problems 4–6, the fractions represent coin values. Rename them as hundredths and then add to find the total value.

(4) $\frac{1}{2} + \frac{1}{20} + \frac{5}{100} + \frac{10}{100} + \frac{1}{100} + \frac{1}{100}$

Hundredths: _____

(5) $\frac{1}{4} + \frac{1}{20} + \frac{1}{10} + \frac{1}{10} + \frac{1}{100} + \frac{1}{2}$

Hundredths: _____

(6) $\frac{1}{4} + \frac{1}{20} + \frac{1}{20} + \frac{1}{20} + \frac{1}{10}$

Hundredths: _____

(7) Write your own coin problem in the space below. Have a partner solve your problem.

_____

_____

_____

Adding Tenths and Hundredths

Use what you know about equivalent fractions to add. Write an equation to show your work.

SRB
166-168

① 2 tenths + 15 hundredths

Equation (in words): _____

② $\frac{68}{100} + \frac{3}{10}$

Equation: _____

③ $\frac{1}{10} + \frac{50}{100}$

Equation: _____

④ $\frac{4}{10} + \frac{60}{100} + \frac{3}{10} + \frac{81}{100}$

Equation: _____

⑤ $1\frac{3}{10} + 5\frac{64}{100}$

Equation: _____

⑥ $3\frac{22}{100} + 2\frac{8}{10}$

Equation: _____

⑦ $\frac{15}{10} + \frac{78}{100}$

Equation: _____

⑧ Nicholas shaded $\frac{40}{100}$ of his hundreds grid. Victor shaded $\frac{5}{10}$ of his grid.

Who shaded more? _____

How much did they shade in all? _____ of a grid

Practice

Write three equivalent fractions.

⑨ $\frac{1}{2} =$ _____

⑩ $\frac{1}{3} =$ _____

⑪ $\frac{1}{4} =$ _____

⑫ $\frac{1}{5} =$ _____

Queen Arlene's Dilemma

① Queen Arlene has a problem. She wants to divide her land among her 3 daughters. She wants her oldest daughter to get $\frac{1}{2}$ of the land and her 2 younger daughters to each get $\frac{1}{3}$ of the land. Can she do it? _____

Use diagrams and words to explain your answer.

② After thinking about it, Queen Arlene decides to keep $\frac{1}{2}$ of her land and have her 3 daughters divide the other $\frac{1}{2}$. She still wants her oldest daughter to get more land than the 2 sisters. Think of a way to use fractions to divide the land.

a. Show a diagram of your answer.

b. Write a fraction addition equation for your answer.

203

Fraction Error Finder

Consider this problem:

A king owns land outside of his castle.

He has partitioned the land to give as gifts to his 5 sons.

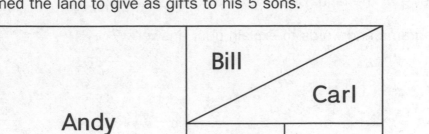

What fraction of the land did the king give to each of his sons?

Here is Zeke's solution:

Andy got $\frac{1}{2}$ *Bill got $\frac{1}{5}$*   *Carl got $\frac{1}{5}$*

Dirk got $\frac{1}{8}$ *Evan got $\frac{1}{8}$*

① Identify Zeke's two errors, correct them, and explain why your answer is correct.

② Write a fraction addition equation to represent the correct answers and show the sum of the pieces of land.

_____

Practice

Use U.S. traditional addition and subtraction.

③ 8,936 + 6,796 = _____   ④ 635 − 392 = _____

⑤ 6,386 + 4,205 = _____   ⑥ 900 − 463 = _____

Subtracting Fractions

Solve the number stories. Use a different strategy for each one.

① Elijah still had $\frac{4}{5}$ of his allowance at the end of the month. Then he spent $\frac{3}{5}$ of his original allowance on a movie ticket and popcorn. How much of Elijah's allowance was left?

a. Fill in the whole box.

Whole

b. Number model with unknown: _____

c. One way to solve a fraction subtraction problem:

d. Answer (with unit): _____

② Kendra's computer battery had $\frac{9}{10}$ of a charge. After her sister Lydia borrowed the computer, the battery had $\frac{3}{10}$ of a charge left. How much of the battery charge did Lydia use?

a. Fill in the whole box.

Whole

b. Number model with unknown: _____

c. Another way to solve a fraction subtraction problem.

d. Answer (with unit): _____

Subtract.

③ $\frac{2}{2} - \frac{1}{2} =$ _____

④ $\frac{11}{6} - \frac{4}{6} =$ _____

⑤ _____ $= 1 - \frac{1}{5}$

Practice

⑥ $8,936 + 6,796 =$ _____

⑦ _____ $= 4,635 - 2,392$

⑧ _____ $- 16,386 + 4,205$

⑨ $65,900 - 48,463 =$

Mixed-Number Subtraction Stories

Solve.

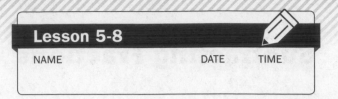

$2\frac{7}{8}$ in.

? $\frac{1}{2}$ in.

(1) Corey drew a line segment $2\frac{7}{8}$ inches long. Then he erased $\frac{1}{2}$ inch. How long is the line segment now?

_____ inches

(2) Mr. Wise needs an $8\frac{2}{5}$-foot strip of wallpaper to finish covering the wall. He has a $10\frac{6}{10}$-foot strip left. How much must he cut off the wallpaper strip to get the amount he needs to finish his task?

_____ feet

(3) The gas tank in Philip's new car holds $16\frac{5}{6}$ gallons of gas. He began his trip to his sister's house with a full tank. If he used $13\frac{2}{3}$ gallons to reach his destination, how much gas is still in the tank?

_____ gallons

(4) The farmer's bag of corn weighed $23\frac{1}{2}$ pounds. After he fed his animals some of the corn, the bag weighed $14\frac{3}{4}$ pounds. How many pounds of corn did he feed to the animals?

_____ pounds of corn

(5) Create a mixed-number subtraction number story using unlike denominators and solve it.

_____

_____

Mixed-Number Subtraction

Solve the number stories. Use a different strategy for each one.

SRB
164-165

(1) The chocolate chip cake recipe calls for $3\frac{1}{3}$ cups of milk. We only have $1\frac{2}{3}$ cups at home. How much more milk do we need?

Whole

    a. Fill in the whole box.

    b. Number model with unknown: _____

    c. One way to solve a mixed-number subtraction problem:

    d. Answer (with unit): _____

(2) Lourdes is listening to an audio book that is 9 hours long. She has listened for $6\frac{1}{6}$ hours so far. How many hours of listening time are left?

Whole

    a. Fill in the whole box.

    b. Number model with unknown: _____

    c. A different way to solve a mixed-number subtraction problem:

    d. Answer (with unit): _____

Subtract. Show your work.

(3) $4\frac{1}{2} - 3\frac{1}{2} =$ _____

(4) _____ $= 5\frac{8}{12} - 5\frac{3}{12}$

(5) $4\frac{2}{5} - 1\frac{4}{5} =$ _____

(6) _____ $= 9\frac{4}{10} - 3\frac{8}{10}$

Practice

(7) _____ $= 54 * 10$

(8) $63 * 100 =$ _____

(9) $86 * 94 =$ _____

(10) $5,715 * 6 =$ _____

Mystery Line Plots

Plot A Unit: _____

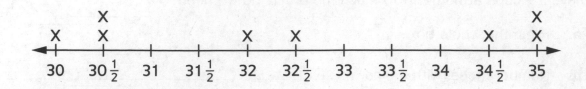

Plot B Unit: _____

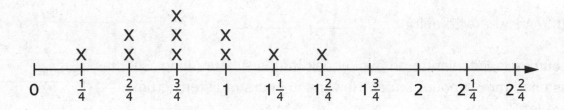

Plot C Unit: _____

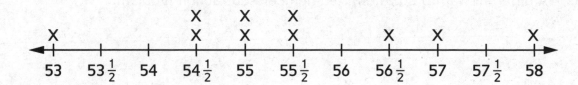

Plot D Unit: _____

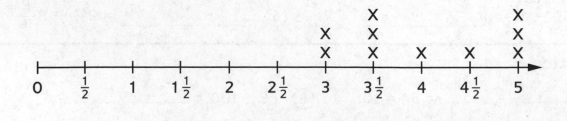

Making a Line Plot

Answer the questions as you make the line plot with your teacher.

① **a.** What is a good title for this line plot? _____

_____

b. Why is this a good title? _____

_____

② **a.** What is a good label for the horizontal axis? _____

_____

b. Why is this a good label? _____

_____

③ What does each stick-on note represent? _____

_____

Use the line plot you created with your teacher to answer the following questions.

④ **a.** What number was rolled the most? _____

b. How many times was this number rolled? _____ times

⑤ **a.** What number was rolled the least? _____

b. How many times was this number rolled? _____ times

Using a Line Plot

Use the line plot you made on *Math Masters*, page TA44 to answer
the questions.

1. Which weight is carried by the most students? _____ pounds

2. How much does the heaviest backpack weigh? _____ pounds

3. How much does the lightest backpack weigh? _____ pounds

4. What is the sum of the weights of the heaviest backpack and
   the lightest backpack?

   Number model with unknown: _____

   Answer: _____ pounds

5. What is the difference between the weights of the heaviest backpack
   and the lightest backpack?

   Number model with unknown: _____

   Answer: _____ pounds

6. What is the difference between the backpack weight carried by
   the most students and the heaviest backpack?

   Number model with unknown: _____

   Answer: _____ pounds

7. What is the difference between the backpack weight carried by
   the most students and the lightest backpack?

   Number model with unknown: _____

   Answer: _____ pounds

Student Growth

Mrs. Welch surveyed her students about how much they had grown over the past year. This is the data she gathered.

Student Growth Over the Past Year (to the nearest $\frac{1}{2}$ inch)	
$1\frac{1}{2}$	$1\frac{1}{2}$
2	$2\frac{1}{2}$
$2\frac{1}{2}$	2
$\frac{1}{2}$	$1\frac{1}{2}$
$2\frac{1}{2}$	$\frac{1}{2}$
1	2
$1\frac{1}{2}$	2
$1\frac{1}{2}$	$\frac{1}{2}$
$3\frac{1}{2}$	$1\frac{1}{2}$
1	1
1	$2\frac{1}{2}$
2	2
$2\frac{1}{2}$	$1\frac{1}{2}$

① Plot the data set on the line plot.

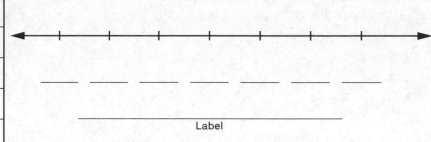

Title

Label

Use the completed line plot to answer the questions.

② What is the greatest number of inches a student grew in a year?

About _____ inch(es)   The least? About _____ inch(es)

③ What is the difference between the greatest and the least number of inches grown?

Number model with unknown: _____   Answer: _____ inch(es)

Practice

Circle the three equivalent fractions in each group.

④ $\frac{1}{4}, \frac{3}{6}, \frac{1}{8}, \frac{2}{8}, \frac{3}{12}$ ⑤ $\frac{3}{4}, \frac{4}{8}, \frac{6}{8}, \frac{5}{6}, \frac{9}{12}$

⑥ $\frac{2}{3}, \frac{1}{5}, \frac{4}{6}, \frac{7}{12}, \frac{8}{12}$ ⑦ $\frac{1}{2}, \frac{5}{10}, \frac{4}{8}, \frac{7}{12}$

Measuring Angles

Math Message

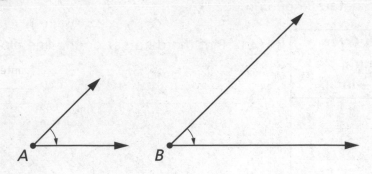

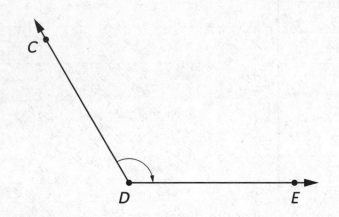

How many whole wedges fit into the angle? _____ wedge(s)

About how much of the fourth wedge is inside the angle? _____ wedge(s)

About how many purple wedges fit? _____ wedge(s)

Rotations

Family Note If your child needs help with the following problems, consider putting up signs in a room in your home to indicate the directions *north, south, east,* and *west.* Do the turns with your child.

Please return this Home Link to school tomorrow.

left turn
counterclockwise

right turn
clockwise

Make the turns described below. Show which way you face after each turn by:

- Drawing a dot on the circle.

- Labeling the dot with a letter.

Example: Face north.

Do a $\frac{1}{2}$ turn counterclockwise.

On the circle, mark the direction you are facing with the letter *A*.

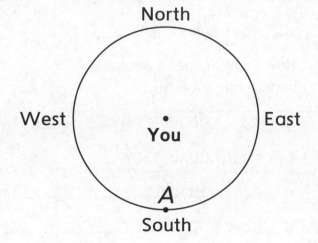

(1) Face north. Do a $\frac{1}{4}$ turn clockwise. Mark the direction you are facing with the letter *B*.

(2) Face north. Do a $\frac{3}{4}$ turn clockwise. Mark the direction you are facing with the letter *C*.

(3) Face east. Do a $\frac{1}{4}$ turn counterclockwise. Mark the direction you are facing with the letter *D*.

(4) Face west. Make less than a $\frac{1}{4}$ turn clockwise. Mark the direction you are facing with the letter *E*.

(5) Face north. Make a clockwise turn that is more than a $\frac{1}{2}$ turn but less than a $\frac{3}{4}$ turn. Mark the direction you are facing with the letter *F*.

(6) Face north. Make a counterclockwise turn that is less than a $\frac{1}{2}$ turn but more than a $\frac{1}{4}$ turn. Mark the direction you are facing with the letter *G*.

Practice

(7) $85 * 50 =$ _____

(8) $416 * 6 =$ _____

(9) _____ $= 597 * 4$

(10) $1,373 * 7 =$ _____

Clock Angles

Use a clock and your knowledge of benchmark angles to help you answer the questions.

1. In one hour, the minute hand on a clock makes a rotation of 360°. How many degrees does it rotate in . . .

   a. 15 minutes? _____

   b. 5 minutes? _____

   c. 30 minutes? _____

   d. 45 minutes? _____

2. How many degrees does the minute hand rotate . . .

   a. from 12:00 to 1:00? _____

   b. from 7:00 to 7:30? _____

   c. from 5:15 to 5:30? _____

   d. from 1:00 to 1:10? _____

   e. from 6:30 to 6:50? _____

   f. from 4:10 to 4:50? _____

3. Explain how you solved Problem 2d. _____

   _____

   _____

   _____

4. Describe two different ways you could solve Problem 2e.

   _____

   _____

   _____

   _____

Estimating Angle Measures

> **Family Note** Our class has been learning about turns, angles, and angle measures. A full turn can be represented by an angle of 360°, a $\frac{1}{2}$ turn by an angle of 180°, a $\frac{1}{4}$ turn by an angle of 90°, and so on. Help your child match the measures below with the angles pictured. (It is not necessary to measure the angles with a protractor.)

Name which angle has the given measure.

Rotation	Degrees
$\frac{1}{4}$ turn	90°
$\frac{1}{2}$ turn	180°
$\frac{3}{4}$ turn	270°
full turn	360°

(1) about 180°   angle _____

(2) about 90°   angle _____

(3) about 270°   angle _____

(4) between 0° and 90°   angle _____

(5) between 90° and 180° angle _____

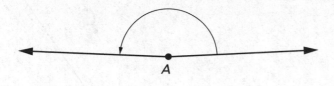

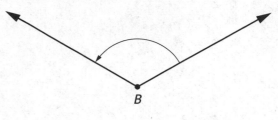

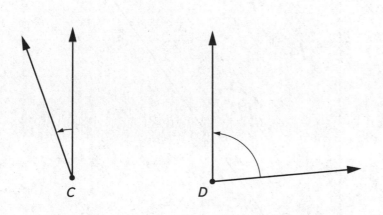

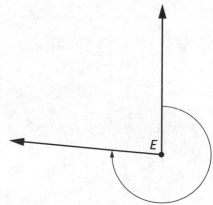

Practice

(6) 5,956 + 4,983 = _____

(7) 60,351 + 86,037 = _____

(8) 41,015 − 517 = _____

(9) 23,730 − 10,769 = _____

Mirror Image

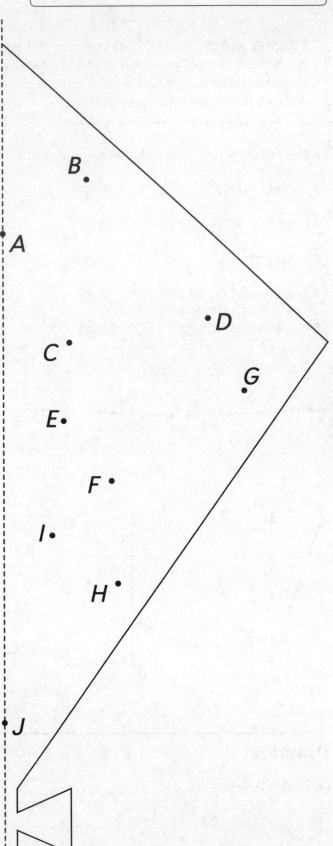

216

Pattern-Block Symmetry Riddles

Use pattern blocks to solve the riddles. Then use your Geometry Template to record your solution to each problem on another piece of paper.

① Build a symmetrical shape using these clues:

- Use exactly 2 red trapezoids and put them together to make a hexagon.

- Use exactly 6 green triangles around the outside of the hexagon.

- Use exactly 8 blocks.

② Build a symmetrical shape using these clues:

- Use exactly 2 red trapezoids.

- Use exactly 5 tan rhombuses.

- Use exactly 7 blocks.

③ Build a symmetrical shape using these clues:

- Build a large triangle.

- Use a yellow hexagon in the center at the bottom of the large triangle.

- Use at least 3 different colors of blocks.

Try This

④ Build a shape that has more than 1 line of symmetry using these clues:

- Use exactly 2 red trapezoids.

- Do not use yellow hexagons.

- The longer sides of the red trapezoids touch and line up together.

- Use a green triangle at the top and at the bottom of the shape.

- Use exactly 10 blocks.

Folding Shapes

Family Note Our class has been studying lines of symmetry—lines that divide figures into mirror images. Help your child look for symmetric shapes in books, newspapers, and magazines, and in objects around the house, such as windows, furniture, dishes, and so on.

Please bring your cutouts to school tomorrow.

(1) Fold a sheet of paper in half. Cut off the folded corner, as shown. Before you unfold the cutoff piece, guess its shape.

SRB
238

   a. Unfold the cutoff piece. What shape is it?

   _____

   b. How many sides of the cutoff
   piece are the same length? _____ sides

   c. How many angles are the same size? _____ angles

   d. The fold is a line of symmetry. Does the
   cutoff piece have any other lines of symmetry? _____

(2) Fold another sheet of paper in half. Fold it in half again. Make a mark on both folded edges 2 inches from the folded corner. Cut off the folded corner. Before you unfold the cutoff piece, guess its shape.

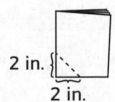

2 in.

2 in.

   a. Unfold the cutoff piece. What shape is it? _____

   b. Are there any other lines of
   symmetry besides the fold lines? _____

   c. On the back of this paper, draw a picture of the
   cutoff shape. Draw all of its lines of symmetry.

Practice

(3) $81 \div$ _____ $= 9$ (4) _____ $\div 9 = 6$

(5) $7 = 42 \div$ _____   (6) _____ $\div 9 = 8$

(7) $36 \div$ _____ $= 4$ (8) $8 =$ _____ $\div 6$

Solving Multistep Number Stories

Fill in the information to solve each number story below.

① Rob's mother has $300 to spend on her son's school clothes. At the store Rob chooses 4 dress shirts at $17 each, 5 T-shirts at $8 each, and 5 pairs of jeans at $16 each. He also wants the sneakers that cost $43 and the dress shoes that cost $55. If his mother buys all these items, how much money will she have left?

To solve the problem, you need to find out the following:

A. Cost for more than one of the same item	B. Total cost of all items	C. Difference between $300 and the total cost
Dress shirts: 4 * 17 = $_____ T-shirts: 5 * 8 = $_____ Jeans: 5 * 16 = $_____	Dress shirts: $_____ T-shirts: $_____ Jeans: $_____ Sneakers: $_____ Dress shoes: $_____ Total: $_____	$3 0 0 − $ [ ] $ [ ]

Solution: Rob's mother will have $_____ left.

② Mrs. Jenson is making some costumes for the play *Snow White and the Seven Dwarfs*. She needs 5 yards of taffeta for the queen and 4 yards of taffeta for Snow White. Taffeta costs $13 per yard. She needs 4 yards of brown denim for each of the 7 dwarfs. Denim costs $8 per yard. The prince's costume will take 5 yards of satin at $10 per yard. A crown for the queen sells for $21. If she has $450 to spend, how much will she have left to purchase ribbons and lace?

A. Total cost for each type of fabric:	B. Total cost:	C. Difference between the amount Mrs. Jenson has and the total cost:
Taffeta: _____ Denim: _____ Satin: _____	_____: $_____ _____: $_____ _____: $_____ _____: $_____ Total: $_____	$ [ ] − $ [ ] $ [ ]

Solution: Mrs. Jenson has $_____ to spend on lace and ribbons.

A Trip to Thrill City

Martha researched the costs for parking, food, and admission to Thrill City Amusement Park. She put the information in a chart for her parents.

Tickets	Purchased Online	Purchased at Park
Parking per vehicle	$22	$25
Meals per person	$13	$19
Admission per person	$49	$64

Solve. Record a long number model with a letter for the unknown quantity and write the answer with the correct unit.

① If Martha, her parents, and 4 friends travel in a van to the park and buy all their tickets at the gate including food, how much will it cost?

Number model with unknown: _____

Answer (with unit): _____

② If all the tickets for the same group are purchased online, how much will they spend?

Number model with unknown: _____

Answer (with unit): _____

③ How much money can they save by buying all the tickets online?

Number model with unknown: _____

Answer (with unit): _____

Expressing Answers to Number Stories

Family Note Today students learned to express solutions to multistep number stories using correct units and single number models. Have your child explain the steps for solving each of the problems below, and then help him or her write these steps as a single number model, including a letter for the unknown quantity.

Solve. Record a long number model with a letter for the unknown quantity and write the answer with the correct unit.

SRB
26, 47,
82-87

① Guillermo hires two painters to paint the walls of his living room. The painters each make $42 an hour for an 8-hour workday. If the work takes 3 days, how much will Guillermo pay the painters?

Number model with unknown: _____

Estimate: _____

Answer (with unit): _____

② Blaine is on vacation in New York City and wants to collect magnets of places he visits to give to all his friends. The Times Square magnets cost $2 each and come in sets of 4. The Statue of Liberty magnets cost $3 each and come in sets of 5. If Blaine buys 12 sets of each type of magnet, how much will he pay?

Number model with unknown: _____

Estimate: _____

Answer (with unit): _____

Practice

③ $45 \div 5 =$ _____

④ $56 \div 8 =$ _____

⑤ $54 \div 9 =$ _____

⑥ _____ $\div 9 = 4$

⑦ _____ $\div 6 = 6$

⑧ _____ $\div 8 = 3$

221

Unit 6: Family Letter

Division; Angles

Division

In Unit 6 your child will divide multidigit numbers using extended division facts, multiples, area models, and partial quotients. Working with more than one division strategy helps students build conceptual knowledge and means that they have more than just one method to choose from. Throughout the unit students solve multistep division number stories involving dividends with multiple digits, learn the meaning of the remainders, and apply their division skills in real-life contexts.

The unit begins with extended division facts. Knowing that $24 \div 4 = 6$ enables students to see that $240 \div 4 = 60$; $240 \div 40 = 6$; $2,400 \div 4 = 600$; and so forth. Students play *Divide and Conquer*, where they practice dividing with extended facts. The confidence they build by working with extended division facts will help them to divide larger numbers with ease.

Students also learn the partial-quotients division method, in which the dividend is divided in a series of steps. The first example below illustrates a model of the partial-quotients method for $1,325 \div 9$. When students partition, or divide, the 1,325 into parts ($900 + 360 + 63 + 2$), it helps them develop their understanding of the algorithm. The second example uses the partial-quotients method. The quotients for each step are added together to give the final answer.

9			
1,325 *s*	$100 * 9 = 900$ 100	1325	9)1,325
		-900	-900 **100**
		425	425
	$40 * 9 = 360$ 40	$+$	-360 **40**
		-360	65
		65	-63 **7**
	$7 * 9 = 63$ 7	-63	2 **147**
	147	2	

Angles

Students continue their work with angle measurement and learn to use both full-circle and half-circle protractors. They learn that angle measurements can be added, and they use this understanding and properties of angles to find unknown angle measures.

Fraction Operations

Students continue working with addition and subtraction of fractions and mixed numbers. They apply their knowledge of multiplication to explore multiplying a fraction by a whole number.

Please keep this Family Letter for reference as your child works through Unit 6.

Vocabulary

Important terms in Unit 6:

complementary angles Angles with measures that equal 90° when added together.

extended division facts Variations of division facts involving multiples of 10, 100, and so on. For example, $720 \div 8 = 90$ is an extended fact related to $72 \div 8 = 9$.

partial quotients A way to divide in which the dividend is divided in a series of steps. The quotients for each step (called partial quotients) are added to give the final answer.

protractor A tool that measures angles in degrees.

reflex angle An angle measure that is between 180° and 360°.

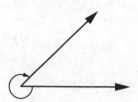

straight angle An angle that measures 180°.

supplementary angles Angles with measures that equal 180° when added together.

Do-Anytime Activities

To work with your child on concepts taught in this unit, try these activities.

1. Practice extended division facts, such as $1,800 \div 30$.

2. Ask your child to help you divide something for dinner into equal portions for each member of your family. For example, ask, "How can we divide the 5 chicken breasts equally for the 4 of us?"

3. Ask questions like these:
   - What kind of angles do you see on a stop sign?
   - What types of angles are on our tile or wood floors, or on the walls?
   - What types of angles are in a rectangular sign?
   - What types of angles do you see in the supports for the bridge?

4. Make up some situations such as those listed directly above, and encourage your child to draw a picture or diagram to show you how to solve it.

Building Skills through Games

In this unit your child will play the following games to increase his or her understanding of division and angles. For detailed instructions on how to play these games, please see the *Student Reference Book.*

Angle Add-Up See *Student Reference Book,* page 248. This game provides practice adding and subtracting angle measures.

Divide and Conquer See *Student Reference Book,* page 254. This game for three players—the Caller, the Brain, and the Calculator—provides practice with extended division facts.

As You Help Your Child with Homework

As your child brings assignments home, it may be helpful to review the instructions together, clarifying them as necessary. The answers listed below will guide you through some of the Home Links in Unit 6.

Home Link 6-1

1. 4; 40

3. a. 5   **b.** 50   **c.** 500   **d.** 5

5. a. 2   **b.** 20   **c.** 200   **d.** 2

7. 2,280   **9.** 6,335

Home Link 6-2

1. Sample answer: $2 * s = 60$; 30 meters

3. 3; 7; 45; 10   **5.** 60   **7.** 60

Home Link 6-3

1. 40, 42, 44, 46, 48, 50;
46 / 2 = b; 23 packages; 46 / 2 = 23

3. 820   **5.** 999

Home Link 6-4

1. Sample estimate: 45 / 3 = 15; 48 ÷ 3 = p; 16 pounds

3. Sample answer: $\frac{3}{6}$; $\frac{4}{8}$

5. Sample answer: $\frac{1}{4}$; $\frac{3}{12}$

Home Link 6-5

1. Sample answer: 115 is the total number of students. 4 is the number of buses. 28 is the number of students per bus. 3 is the number of students left over after dividing evenly.

2. Sample answer: Because 28 students from each class can be on a bus and there are 3 students left over, 3 buses will have 29 students. Then, because each bus needs a teacher, 3 buses will have 30 passengers on them and 1 bus will have 29 passengers.

Mr. Atkins's class has too many students to fit on one bus. So he can go on the bus with most of his students, and 2 students will have to ride on another bus. His bus will have 30 passengers.

Mrs. Gonzales's class has the fewest students. Because she has 27 students and adding herself makes 28 passengers, her bus will have room for Mr. Atkins's 2 extra students.

Mr. Bates and his students are a perfect fit for a bus. There will be 30 passengers on his bus.

Ms. Smith and her students fit on a bus, with room for one more. However, that spot is not needed.

3. $\frac{7}{8}$ **5.** $\frac{2}{5}$

Home Link 6-6

1. 12,000; 7; 16,000, 11   **3.** 8,000 pounds

5. $\frac{7}{8}$ **7.** $\frac{53}{100}$

Home Link 6-7

1. Sample answer:

$$5\overline{)360}$$
$$-\;350 \quad | \quad 70$$
$$\overline{10}$$
$$-\;10 \quad | \quad 2$$
$$\overline{0 \quad\;\; 72}$$

Sample estimate: 350 ÷ 5 = 70; 360 ÷ 5 = p; 72 prizes; 0 prizes

3. Sample estimate: 160 / 8 = 20; 23

5. 0.08, 0.34, 0.98, 9.8   **7.** >

Home Link 6-8

1.
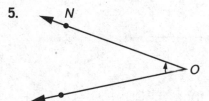

Sample number models are given.

$124 / 8 = s$; $15\frac{4}{8}$, or $15\frac{1}{2}$ strawberries;
$124 / 8 \rightarrow 15$ R4;

B. Reported it as a fraction;
Sample answer: You can cut the remaining strawberries into halves.

3. $\frac{3}{8}, \frac{3}{6}, \frac{3}{5}, \frac{3}{3}$ 5. $\frac{1}{2}, \frac{2}{3}, \frac{6}{8}, \frac{99}{100}$

Home Link 6-9

1. Right; 90°   3. Acute; 45°

5. 692   7. 680

Home Link 6-10

1. 60°   3. 84°

5.

7. 65,811   9. 64,091

Home Link 6-11

1. Sample answer: $30° + y = 90°$; 60°

3. Sample answer: $90° - z = 75°$; 15°

5. Sample answer: $180° - 60° = a$; 120°

7. $\frac{7}{12}, \frac{7}{10}, \frac{7}{9}, \frac{7}{8}$

Home Link 6-12

1. **a.** Strawberries; $\frac{3}{12} + \frac{1}{12} = b$; $\frac{4}{12}$ pound
   b. $\frac{3}{12} - \frac{1}{12} = p$; $\frac{2}{12}$ pound

3. $4\frac{2}{8} + 1\frac{3}{8} = p$; $5\frac{5}{8}$ pounds

5. 2,400

Home Link 6-13

1. 45 children; Sample answer:

X X X	X X X	X X X	X X X	X X X
X X X	X X X	X X X	X X X	X X X
X X X	X X X	X X X	X X X	X X X

5 groups of 9;
$9 + 9 + 9 + 9 + 9 = 45$; $5 * 9 = 45$

3. $2\frac{2}{5}$ veggie pizzas; Sample answer:

4 groups of $\frac{3}{5}$;
$\frac{3}{5} + \frac{3}{5} + \frac{3}{5} + \frac{3}{5} = \frac{12}{5}$; $4 * \frac{3}{5} = \frac{12}{5}$

5. 19

More Extended Facts Practice

Solve.

①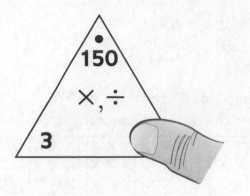

Extended fact: 150 ÷ 3 = _____

②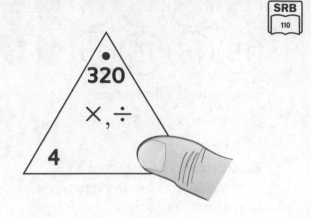

Extended fact: 320 ÷ 4 = _____

③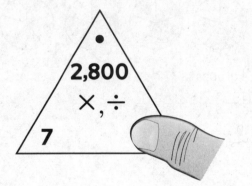

Extended fact: 2,800 ÷ 7 = _____

④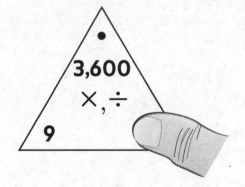

Extended fact: 3,600 ÷ 9 = _____

⑤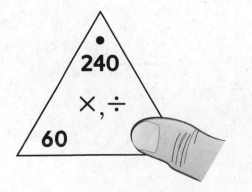

Extended fact: 240 ÷ 60 = _____

⑥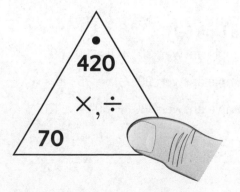

Extended fact: 420 ÷ 70 = _____

Solving Extended Division Facts

Write a basic division fact and an extended division fact for each Fact Triangle.

SRB
110

①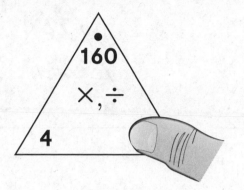

Basic fact: $16 \div 4 =$ _____

Extended fact: $160 \div 4 =$ _____

②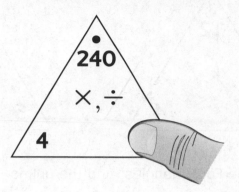

Basic fact: _____

Extended fact: _____

Solve.

③   **a.** $25 \div 5$ _____

    b. $250 \div 5 =$ _____

    c. $2{,}500 \div 5 =$ _____

    d. $250 \div 50 =$ _____

④   **a.** $36 \div 4 =$ _____

    b. $360 \div 4 =$ _____

    c. $3{,}600 \div 4 =$ _____

    d. $360 \div 40 =$ _____

⑤   **a.** $18 \div 9 =$ _____

    b. $180 \div 9 =$ _____

    c. $1{,}800 \div 9 =$ _____

    d. $180 \div 90 =$ _____

⑥   **a.** $42 / 7 =$ _____

    b. $420 / 7 =$ _____

    c. $4{,}200 / 7 =$ _____

    d. $420 / 70 =$ _____

Practice

⑦ $456 * 5 =$ _____

⑧ $720 * 8 =$ _____

⑨ $905 * 7 =$ _____

Unknown Factors

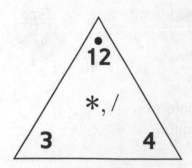
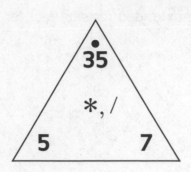

Using the Fact Triangles, find the unknown factors.

(1) 3 * _____ = 12

(2) _____ * 3 = 12

(3) 4 * _____ = 12

(4) _____ * 4 = 12

(5) 5 * _____ = 350

(6) _____ * 50 = 350

(7) 7 * _____ = 350

(8) _____ * 70 = 350

(9) 720 = 90 * _____

(10) 720 = 9 * _____

(11) 720 = 8 * _____

(12) 720 = 80 * _____

Solve.

(13) 6 * _____ = 540

(14) 420 = 7 * _____

Finding Garden Plot Dimensions

Mrs. Berkman is putting up string fences to create 8 rectangular garden plots for families as part of a community garden along the path by the river. Some families gave her the area and the length of one side for their plot, and others gave her the lengths of both sides of the plots.

- Help her plan where to put the string by filling in the unknown information below.
- Then answer the questions and draw a diagram for the community garden according to the directions given.

Community Garden Plot Dimensions			
Family Name for Garden Plot	Area in Square Feet	Side A (ft)	Side B (ft)
Alvarez	72	8	
Segal	36		4
Wong		7	6
Sharp	40		8
Edwards		9	9
Fine	45		5
Girard	56	8	
Harrison	48	6	

① How many square feet are in all of the family plots combined? _____ square feet

② Mrs. Berkman plans to put the garden plots 3 feet apart, with one side of each plot running parallel to the river path. How long would the river path be from the beginning to the end of the community garden? Draw a rough sketch of the garden plots below or on the back of this page, and label only the side facing the river path and the 3-foot spaces to show how you got your answer.

Answer: _____

A Question about Rectangles

Record all of the possible whole-number combinations of side measures for rectangles with a perimeter of 28 units.

SRB
200, 204

Equations for Perimeter of Rectangle	Perimeter	Equations for Area of Rectangle	Area
$1 + 1 + 13 + 13 = p$	28 units	$1 * 13 = A$	13 sq units
	28 units		_____ sq units
	28 units		_____ sq units
	28 units		_____ sq units
	28 units		_____ sq units
	28 units		_____ sq units
	28 units		_____ sq units

Finding the Unknown Side Length

Solve.

(1)

	s
2 meters	60 square meters

How long is the unknown side s?

Equation with unknown: _____

Answer: _____ meters

(2)

t

6 meters | 420 square meters

What is the length of the unknown side t?

Equation with unknown: _____

Answer: _____ meters

(3) Fill in the unknown information about some rectangular rooms in a museum.

Room	Length in Yards	Width in Yards	Area in Square Yards
A	6		18
B		8	56
C	9	5	
D		9	90

(4) A store is rectangular in shape with an area of 2,700 square feet. It has a length of 90 feet. How wide is it?

Equation with unknown: _____

Answer: _____ feet

Practice

(5) $420 \div 7 =$ _____   **(6)** _____ $= 3{,}600 / 6$ **(7)** $5{,}400 \div 90 =$ _____

Solving Division Number Stories

Fill in the lists of multiples to help you, if needed.

SRB
55, 110, 114

(1) Rosario sells bicycle wheels in packages of 2. If a store orders 46 wheels, how many packages will she send?

20 [2s] = _____   Number model with unknown: _____

21 [2s] = _____   Answer: _____ packages

22 [2s] = _____   Number model with answer: _____

23 [2s] = _____

24 [2s] = _____

25 [2s] = _____

(2) Doug is placing apples in bags for a picnic. He can fit 6 apples in a bag. If he has 92 apples, how many bags will he need?

10 [6s] = _____   Number model with unknown:_____

11 [6s] = _____   Answer: _____ bags

12 [6s] = _____   Number model with answer: _____

13 [6s] = _____

14 [6s] = _____

15 [6s] = _____

16 [6s] = _____

17 [6s] = _____

18 [6s] = _____

Practice

(3) $82 * 10 =$ _____   **(4)** _____ $= 25 * 30$ **(5)** $333 * 3 =$ _____

Partial-Quotients Division

Family Note In this lesson students are introduced to the partial-quotients method, in which a number is divided in a series of steps. The quotients for each step (called partial quotients) are added to give the final answer. For example, to divide 96 by 6, students use extended multiplication facts such as 6 * **10** = 60 to find the partial quotient. Then with the remaining 36, they use an "easy" multiplication fact, 6 * **6**, to finish solving the problem. These two partial quotients are added together, 10 + 6, to find the exact quotient of 16. So 96 ÷ 6 = 16.

Estimate. Write a number model with an unknown to represent the problem. Then solve using partial quotients.

(1) Jordan has 3 Great Dane puppies. At 6 weeks old, their combined weight is 48 pounds. Assuming that they all weigh about the same amount, how much does each puppy weigh?

Estimate: _____

Number model with unknown: _____

Answer: _____ pound(s)

(2) Four sisters love barrettes. They have a value pack that contains 92 barrettes. How many barrettes can each sister have if they share equally?

Estimate: _____

Number model with unknown _____

Answer: _____ barrette(s)

Practice

Name two equivalent fractions for each fraction given.

(3) $\frac{1}{2}$ ____ ____

(4) $\frac{1}{3}$ ____ ____

(5) $\frac{25}{100}$ ____ ____

(6) $\frac{6}{8}$ ____ ____

Fruit Baskets

The fourth-grade chess team is planning a fundraiser. They are going to sell fruit baskets. Oscar is in charge of oranges for the baskets. Three students brought oranges. Olivia brought 29 oranges, Ozzie brought 31 oranges, and Olga brought 27 oranges.

1. Each basket must have at least 5 oranges. Some baskets may have 6 oranges if there are any extras after each basket has 5 oranges.

   a. How many baskets will be needed for the oranges? Show or explain your thinking.

   b. How many baskets will have 6 oranges and how many will have 5 oranges?

Fruit Baskets (continued)

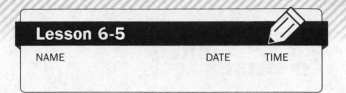

(2) The baskets will be put into boxes for delivery. Each box can hold at most 7 baskets. What is the least number of boxes needed for the fundraiser? Show or explain your thinking.

(3) Explain how you used remainders in two different ways to help you answer Problems 1 and 2.

235

Assigning People to Buses

Mr. Atkins is organizing the 4th- and 5th-grade field trip to the science museum. He asked his students to help him figure out which students and teachers should go on each bus. The number of students in each class is shown in the table below:

SRB
113-115

Mr. Atkins's 4th-grade class	31 students
Ms. Smith's 4th-grade class	28 students
Mr. Bates's 5th-grade class	29 students
Mrs. Gonzales's 5th-grade class	27 students

Important information:

• 4 buses have been ordered.
• The maximum number of passengers is 30 per bus.
• Each bus must have 1 teacher.

Cary said he solved the problem this way:

115 / 4 is 28 with a remainder of 3.

① What do the numbers in his sentence mean?

② Which students and teachers should go on each bus? Explain why.

Practice

③ $\frac{3}{8} + \frac{4}{8} =$ _____   ④ $\frac{5}{6} + \frac{3}{6} =$ _____   ⑤ $\frac{4}{5} - \frac{2}{5} =$ _____   ⑥ $\frac{7}{10} - \frac{3}{10} =$ _____

Record-Setting Food Weights

Use the information in the table to solve the problems about record-setting food weights.

Food	Weight	Date Weighed	Location
Apple	4 pounds, 1 ounce	October 2005	Hirosaki, Japan
Bagel	868 pounds	August 2004	Syracuse, New York
Burrito	12,785 pounds	November 2012	La Paz, Mexico
Cheesecake	6,900 pounds	September 2013	Lowville, New York
Lemon	11 pounds, $9\frac{7}{10}$ ounces	January 2003	Kfar Zeitim, Israel
Onion	18 pounds, 1 ounce	September 2012	Harrogate, England
Pizza	26,883 pounds	December 1990	Norwood, South Africa
Pumpkin	2,009 pounds	September 2012	Topsfield, Massachusetts
Sweet Potato	81 pounds, 9 ounces	March 2004	Lanzarote, Spain
Watermelon	268 pounds, 13 ounces	September 2005	Hope, Arkansas

① About how many ounces would two of the record-setting apples weigh?

   About _____ ounce(s)

② About how many tons does the pumpkin weigh? About _____ ton(s)

③ About how many tons does the burrito weigh? About _____ ton(s)

④ The sweet potato weighs about _____ times more than the apple.

⑤ A *kilogram* is a little more than 2 pounds. Which food weighs about 1,000 kilograms?

   _____ Which foods weigh more than 1,000 kilograms?

   _____

⑥ Use data from the table to write and solve your own problem on the back of this page.

Converting Units of Weight

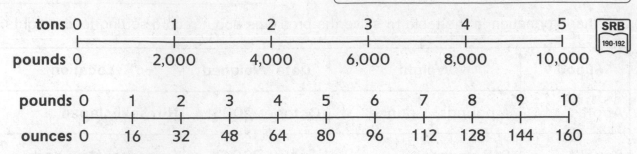

tons	0	1	2	3	4	5
pounds	0	2,000	4,000	6,000	8,000	10,000

pounds	0	1	2	3	4	5	6	7	8	9	10
ounces	0	16	32	48	64	80	96	112	128	144	160

SRB 190-192

Use the measurement scales to help you solve the problems.

①

Tons	Pounds
1	2,000
6	
	14,000
8	
	22,000

②

Pounds	Ounces
1	16
5	
9	
	160
15	

③ The army chef is ordering food for the troops. She ordered 2 tons of rice, 1 ton of pasta, and 1 ton of flour. How many pounds of food did she order?

Answer: _____ pound(s)

④ Potatoes come in 8-pound bags. How many ounces do 12 bags weigh?

Answer: _____ ounce(s)

Practice

⑤ $\frac{4}{8} + \frac{3}{8} =$ _____ ⑥ _____ $= \frac{5}{8} - \frac{3}{8}$ ⑦ _____ $= \frac{5}{10} + \frac{3}{100}$ ⑧ $\frac{60}{100} + \frac{4}{10} =$ _____

Partial Quotients

Estimate. Write a number model to represent the problem. Solve using partial quotients. **SRB** 47, 113-114

(1) The carnival committee has 360 small prizes to distribute to 5 booths. How many prizes will each booth get?

Estimate: _____

Number model with unknown:

_____

(2) The mall needs a row of parking spaces. The length of the parking area is 2,711 feet. If each parking space is 9 feet wide, how many spaces will there be?

Estimate: _____

Number model with unknown:

_____

Answer: _____ prizes

How many prizes are left over? ___ prizes

Answer: _____ spaces

How many feet are left over? ___ feet

Solve using partial quotients. Show your work on the back of this page.

(3) 161 / 7 Estimate: _____   Answer: _____

(4) 576 / 4 Estimate: _____   Answer: _____

Practice

Put these decimals in order from least to greatest.

(5) 0.98, 0.34, 9.8, 0.08 _____, _____, _____, _____

(6) 0.11, 0.01, 0.10, 1.0 _____, _____, _____, _____

Use <, >, or = to compare the decimals.

(7) 0.65 _____ 0.5   (8) 37.9 _____ 37.96

A Remainder of One

Use 25 centimeter cubes to represent the 25 ants in the story *A Remainder of One*.

① Divide the cubes into 2 equal rows. Draw what you did.

Cubes per row? _____ cube(s)

Cubes left over? _____ cube(s)

Number model with unknown:

_____

② Divide the cubes into 3 equal rows. Draw what you did.

Cubes per row? _____ cube(s)

Cubes left over? _____ cube(s)

Number model with unknown:

_____

③ Divide the cubes into 4 equal rows. Draw what you did.

Cubes per row? _____ cube(s)

Cubes left over? _____ cube(s)

Number model with unknown:

_____

④ Divide the cubes into 5 equal rows. Draw what you did.

Cubes per row? _____ cube(s)

Cubes left over? _____ cube(s)

Number model with unknown:

_____

Interpreting Remainders

① Mrs. Patel brought a box of 124 strawberries to the party. She wants to divide the strawberries evenly among 8 people. How many strawberries will each person get?

② Mr. Chew has a box of 250 pens. He asks Maurice to divide the pens into groups of 8. How many groups can Maurice make?

SRB
47,
113-116

Number model with unknown:

_____

Answer:

_____ strawberries

Number model with answer:

_____

What did you do about the remainder? Circle the answer.

A. Ignored it

B. Reported it as a fraction

C. Rounded the answer up

Why? _____

_____

_____

Number model with unknown:

_____

Answer:

_____ groups

Number model with answer:

_____

What did you do about the remainder? Circle the answer.

A. Ignored it

B. Reported it as a fraction

C. Rounded the answer up

Why? _____

_____

_____

Practice

Order the fractions from smallest to largest.

③ $\frac{3}{6}, \frac{3}{3}, \frac{3}{5}, \frac{3}{8}$ _____, _____, _____, _____

④ $\frac{1}{4}, \frac{1}{8}, \frac{1}{2}, \frac{1}{5}$ _____, _____, _____, _____

⑤ $\frac{2}{3}, \frac{1}{2}, \frac{6}{8}, \frac{99}{100}$ _____, _____, _____, _____

⑥ $\frac{4}{5}, \frac{81}{100}, \frac{4}{6}, \frac{2}{10}$ _____, _____, _____, _____

241

Measuring Angles

Cut out the angle measurer and use a pencil to poke a hole through the center.

Label each angle *acute*, *right*, or *obtuse*.

Then use the angle measurer to measure each angle.

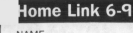
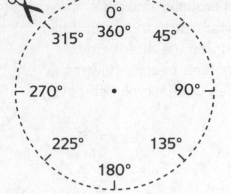

SRB
207,
228-229

①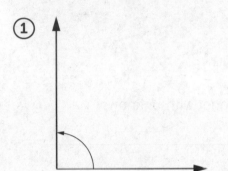

Type of angle: _____

Angle measure: _____

②

Type of angle: _____

Angle measure: _____

③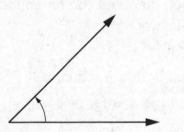

Type of angle: _____

Angle measure: _____

④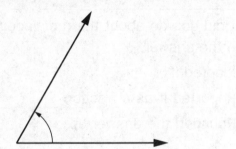

Type of angle: _____

Angle measure: _____

Practice

Multiply.

⑤
```
   1 7 3
*      4
```

⑥
```
   2 4 7
*      6
```

⑦
```
   3 4
* 2 0
```

Measuring Angles with a Protractor

First estimate whether the angles measure more or less than 90°. Then use a half-circle protractor to measure them.

SRB
208-210

(1) ∠A: _____°

(2) ∠B: _____°

(3) ∠C: _____°

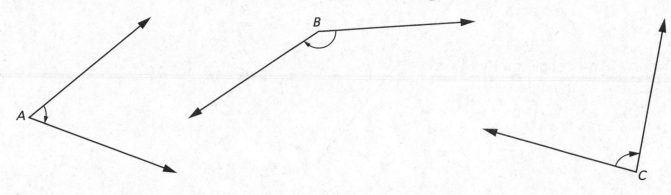

(4) ∠QRS: _____°

(5) ∠NOP: _____°

(6) ∠KLM: _____°

Practice

(7)
```
   2 3, 8 0 7
+  4 2, 0 0 4
_____
```

(8)
```
   5 3 0, 0 8 3
+  2 8 3, 6 9 0
_____
```

(9)
```
   8 7, 9 4 2
−  2 3, 8 5 1
_____
```

(10)
```
   6 0 0, 2 9 9
−  5 1 0, 3 4 5
_____
```

243

Math Message

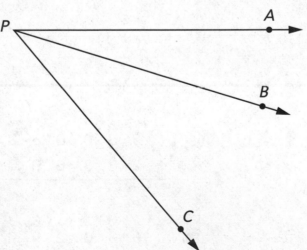

The measure of angle *APB* is 17°. The measure of angle *BPC* is 33°. Without using a protractor, find the measure of another angle in the diagram.

Answer: Angle _____ measures _____ °.

Math Message

The measure of angle *APB* is 17°. The measure of angle *BPC* is 33°. Without using a protractor, find the measure of another angle in the diagram.

Answer: Angle _____ measures _____ °.

244

Putting Together Angles

Use a half-circle protractor to measure each large angle.

(1) ∠ACB: _____°

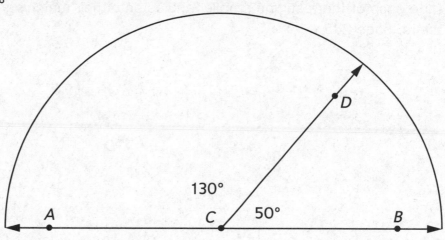

130°   50°

A   C   B   D

(2) ∠EFG: _____°

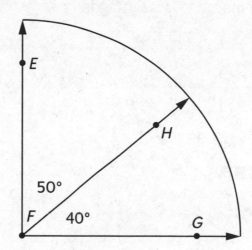

E

H

50°

F   40°   G

(3) ∠IJK: _____°

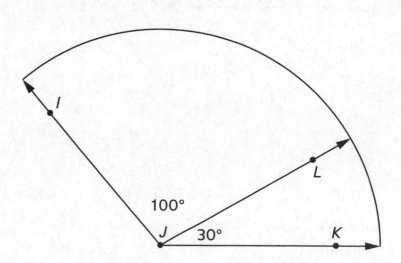

I

L

100°

J   30°   K

245

Putting Together Angles (continued)

Cut out the three large angles. Cut each large angle in two along the dotted line. Glue or tape each of the resulting smaller angles onto their matching angles on *Math Masters*, page 245.

④

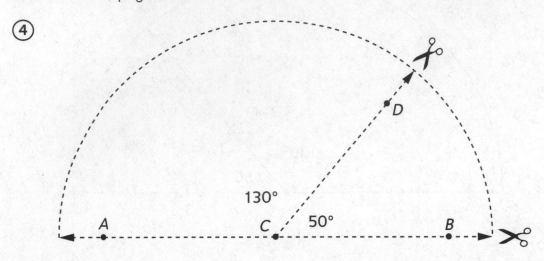

130°   50°

A   C   B

⑤

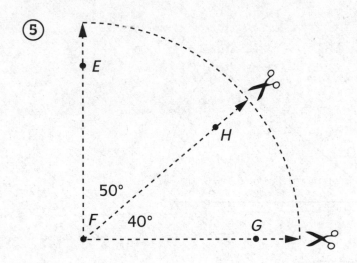

E

H

50°

F   40°   G

⑥

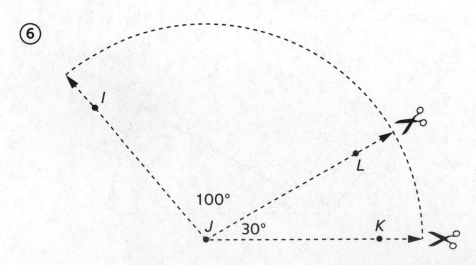

I

L

100°

J   30°   K

Combining Angles

Use the angles to answer the questions.

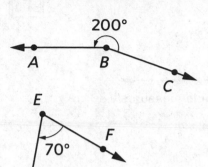

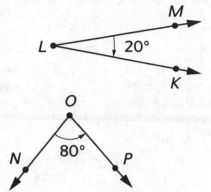

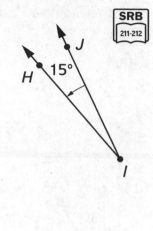

SRB
211-212

1. Which angles could be combined to form a right angle?

   Answer: ∠ _____ and ∠ _____

   Equation: _____ ° + _____ ° = _____ °

2. Which angles could be combined to form an angle measuring 150°?

   Answer: ∠ _____ and ∠ _____

   Equation: _____ ° + _____ ° = _____ °

3. Which angles could be combined to form an angle measuring 95°?

   Answer: ∠ _____ and ∠ _____

   Equation: _____ ° + _____ ° = _____ °

4. Which angles could be combined to form an angle measuring 270°?

   Answer: ∠ _____ and ∠ _____

   Equation: _____ ° + _____ ° = _____ °

5. a. How many different acute angles can you make by combining two of the angles?

      _____ acute angles

   b. List the combinations of angles. _____

   c. Use <, >, or = to write number sentences comparing the sums of the measures
      of the pairs of angles you listed in Problem 5b to 90°.

   _____

Finding Angle Measures

Find the unknown angle measures in Problems 1–6. Do *not* use a protractor.

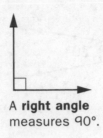

A **right angle** measures 90°.

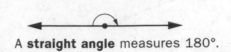

A **straight angle** measures 180°.

①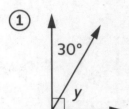

Equation with unknown: _____

Answer: _____

②

Equation with unknown: _____

Answer: _____

③

Equation with unknown: _____

Answer: _____

④

Equation with unknown: _____

Answer: _____

⑤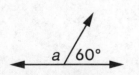

Equation with unknown: _____

Answer: _____

⑥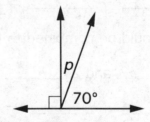

Equation with unknown: _____

Answer: _____

Practice

Order the fractions from smallest to largest.

⑦ $\frac{7}{10}, \frac{7}{8}, \frac{7}{12}, \frac{7}{9}$ _____

⑧ $\frac{5}{9}, \frac{99}{100}, \frac{1}{4}, \frac{9}{10}$ _____

248

Solving Number Stories with Unlike Denominators

Solve these problems about a vacation. Show your work under each problem.

① Kenyon's suitcase weighed $51\frac{3}{4}$ pounds, but the airline weight limit is 50 pounds. If Kenyon removed an object weighing $2\frac{1}{2}$ pounds, what did his suitcase weigh then?

_____ pounds

How much less than the limit does his suitcase weigh now?

Answer: _____ pound

② Carl's suitcase was 3 pounds over the limit. Carl transferred some items to his carry-on bag: a jacket ($\frac{3}{4}$ pound); a paperback book ($\frac{1}{2}$ pound); a camera ($1\frac{3}{4}$ pounds); and a cap ($\frac{1}{4}$ pound). How much did Carl's suitcase weigh then?

_____ pounds

Which object could he return to his suitcase and still meet the weight limit of 50 pounds?

_____

③ During the flight, Carl began to read a book that had 100 pages. He read $\frac{2}{10}$ of the book before taking a nap and $\frac{40}{100}$ more when he woke up. How much of the book is left for Carl to read?

_____ of the book

④ Kenyon budgeted about $\frac{6}{100}$ of his spending money for souvenirs and $\frac{4}{10}$ of his money for food on his vacation. How much of his budget is left for other things?

_____ of his budget

⑤ Kenyon paid the entire $20 fare for the taxi he shared with Carl to the hotel. Carl repaid $\frac{5}{10}$ of his share of half of the taxi fare after lunch and another $\frac{30}{100}$ of his share after supper. How much of his share of the taxi ride does Carl still owe Kenyon?

_____ of his share

⑥ Carl and Kenyon were at the museum for $5\frac{1}{4}$ hours. They spent $\frac{1}{4}$ of an hour in the gem room, $\frac{1}{2}$ hour in the archeology section, $\frac{3}{4}$ of an hour in the theater, and $\frac{1}{2}$ hour eating in the cafeteria. How much time did they spend at the other museum exhibits?

_____ hours

Solving Fish Number Stories

Solve these problems about a fishing contest. Show your work under each problem.

① Last year's heaviest fish weighed $17\frac{3}{8}$ pounds. This year's weighed $19\frac{1}{8}$ pounds. How much more did this year's heaviest fish weigh than last year's?

Number model with unknown: _____

Answer: _____ pound(s)

② The longest fish this year measured $23\frac{1}{3}$ inches. This was $2\frac{2}{3}$ inches longer than the longest fish from last year. How long was the longest fish from last year?

Number model with unknown: _____

Answer: _____ inch(es)

③ The 3 heaviest fish together weighed $55\frac{5}{8}$ pounds. The heaviest fish weighed $19\frac{1}{8}$ pounds and the second heaviest weighed $18\frac{3}{8}$ pounds. How much did the third heaviest fish weigh?

Number model with unknown: _____

Answer: _____ pound(s)

Try This

④ The 3 winners each got a trophy. The total weight of the 3 trophies is $41\frac{1}{4}$ pounds. If each trophy weighed the same amount, how much did each trophy weigh?

Number model with unknowns: _____

Answer: _____ pound(s) each

Solving Number Stories

Write a number model with an unknown to represent each problem. Then solve.

SRB
47,
162-165

(1) Martin had some leftover fruit from making fruit salad. He had $\frac{3}{12}$ pound of strawberries and $\frac{1}{12}$ pound of blueberries.

Which fruit weighed more? _____

a. How many pounds of fruit did Martin have left?

Number model with unknown: _____

Answer: _____ pound

b. How much more did the strawberries weigh than the blueberries?

Number model with unknown: _____

Answer: _____ pound

(2) Charlotte and Beth each made a vegetable salad to take to a reunion. Together the salads weighed 6 pounds. Charlotte's salad weighed $3\frac{1}{2}$ pounds.

a. How much did Beth's salad weigh?

Number model with unknown: _____

Answer: _____ pounds

b. How much more did Charlotte's salad weigh than Beth's?

Number model with unknown: _____

Answer: _____ pound

(3) Andy's potato salad weighed $1\frac{3}{8}$ pounds more than Mardi's. Mardi's potato salad weighed $4\frac{2}{8}$ pounds. How much did Andy's potato salad weigh?

Number model with unknown: _____

Answer: _____ pounds

Practice

(4) $826 * 5 =$ _____

(5) $48 * 50 =$ _____

Solving Equal-Groups Number Stories

Solve. Use drawings, words, and equations to represent the problems.

1. Sherise bought 2 bags of oranges. Each bag has 9 oranges. How many oranges

   does she have? _____ oranges

   Drawing:

   Words: ____ groups of ____

   Addition equation: ____ + ____ = _____

   Multiplication equation: ____ * ____ = _____

bags	oranges per bag	total oranges

2. Tyler has 3 bunches of bananas. Each bunch has 4 bananas. How many bananas

   does he have? _____ bananas

   Drawing:

   Words: ____ groups of ____

   Addition equation: ____ + ____ + ____ = _____

   Multiplication equation: ____ * ____ = _____

bunches	bananas per bunch	total bananas

3. Noah has 5 small bunches of grapes. Each bunch has 6 grapes. How many grapes

   does he have? _____ grapes

   Drawing:

   Words: ____ groups of ____

   Addition equation: ____ + ____ + ____ + ____

   + ____ = _____

   Multiplication equation: ____ * ____ = _____

bunches	grapes per bunch	total grapes

Solving Missing-Groups Number Stories

① Murphy eats $\frac{1}{2}$ pound of dog food per day. How many days will it take him to eat 5 pounds?

Drawing: Words: _____ days of $\frac{1}{2}$ pound = 5 pounds

Multiplication equation: _____ $* \frac{1}{2} = 5$

Number of Days	Pounds Eaten Per Day	Number of Pounds
	$\frac{1}{2}$	5

Answer: _____ days

② Lulu eats $\frac{5}{8}$ pound of food a day. How many days will it take her to eat a 5-pound bag of food?

Drawing: Words: _____ days of $\frac{5}{8}$ pound = 5 pounds

Multiplication equation: _____ $* \frac{5}{8} = 5$

Number of Days	Pounds Eaten Per Day	Number of Pounds
	$\frac{5}{8}$	5

Answer: _____ days

③ Bo eats $\frac{3}{4}$ pound of food a day. How many days will it take him to eat 6 pounds of food?

Drawing: Words: _____ days of $\frac{3}{4}$ pound = 6 pounds

Multiplication equation: _____ $* \frac{3}{4} = 6$

Number of Days	Pounds Eaten Per Day	Number of Pounds
	$\frac{3}{4}$	6

Answer: _____ days

Multiplying a Fraction by a Whole Number

Write a number story to go with each set of equations.

(1) Addition equation with unknown: $\frac{1}{2} + \frac{1}{2} + \frac{1}{2} + \frac{1}{2} + \frac{1}{2} = r$

Multiplication equation with unknown: _____

Number story: _____

_____

_____

Answer: _____

Drawing: Fill in the blanks.

		Total
	$\frac{1}{2}$	

(2) Multiplication equation with unknown: $4 * \frac{3}{4} = t$

Addition equation with unknown: _____

Number story: _____

_____

_____

Answer: _____

Drawing: Fill in the blanks.

		Total
4		

Multiplying a Fraction by a Whole Number

Solve. Use drawings, words, and equations to represent the problems.

(1) 5 vans were needed for a camp field trip. There were 9 children per van.

How many children went on the field trip? _____ children

Drawing: Words: _____ groups of _____

                                Addition equation: _____

                                Multiplication equation: _____

(2) Penny and her 2 friends each ate $\frac{1}{6}$ of a cake. How much cake did they eat?

_____ of a cake

Drawing: Words: _____ groups of _____

                                Addition equation: _____

                                Multiplication equation: _____

(3) Christopher wants to give his 4 friends $\frac{3}{5}$ of a veggie pizza each.

How much veggie pizza will he need? _____ veggie pizzas

Drawing: Words: _____ groups of _____

                                Addition equation: _____

                                Multiplication equation: _____

Practice

(4) $84 / 6 =$ _____   **(5)** $76 \div 4 =$ _____   **(6)** _____ $= 90 \div 5$

Multiplication of a Fraction by a Whole Number; Measurement

Fractions

Unit 7 begins with students applying and extending their previous understandings of multiplying whole numbers to multiplying a fraction by a whole number. Your child will multiply fractions by whole numbers in different ways: using concrete objects, drawing pictures, and writing equations. Using a variety of strategies helps students build conceptual knowledge and gives them more than one method to choose from when solving problems.

Consider this number story, for example: *Mattie needs $\frac{1}{2}$ cup of granola for each member of her family. She has 5 family members. How much granola does she need for everyone in the family?*

Below are examples of different strategies students might use to solve the problem.

- Use repeated addition: $\frac{1}{2} + \frac{1}{2} + \frac{1}{2} + \frac{1}{2} + \frac{1}{2} = \frac{5}{2}$, or $2\frac{1}{2}$ cups of granola

- Apply relational thinking: Two $\frac{1}{2}$s are 1. Four $\frac{1}{2}$s are two. Another $\frac{1}{2}$ is $2\frac{1}{2}$.

- Draw a picture:

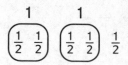

- Use fraction circles:

- Use equations: $5 * \frac{1}{2} = \frac{(5 * 1)}{2} = \frac{5}{2}$

In this unit students create drawings or use models, such as fraction circles or fraction strips, to explain their thinking as they apply their skills in real-life contexts involving time, weight, capacity, and money.

Measurement

In Unit 7 students work with increasingly complex measurement conversion problems. They explore U.S. customary units of capacity, including the cup, pint, quart, and gallon, and solve number stories involving conversions between whole numbers and fractions of units.

Students also convert between pounds and ounces in the course of solving real-world number stories involving U.S. customary units of weight. Lesson 7-12 challenges students with number stories involving decimals in a money context. Here they apply their understanding of fraction/decimal equivalencies and fraction operations to solve the problems. According to the Common Core State Standards, students are not expected to perform operations with decimals until fifth grade. However, the link established through these activities between different representations of numbers, especially fractions and decimals, is a key prerequisite concept for success with decimal computation. Problems like the ones presented in this unit build the foundation for that later work.

Line Plots

Line plots are used to organize and display data. Students analyze data measured to an eighth of a unit, create their own line plots, and use line plots to solve problems involving computations with fractions and mixed numbers.

Division

Students estimate, solve, and assess the reasonableness of answers to multistep division number stories. They plan strategies and write number models with letters for the unknown quantities, explaining how they found each answer and checking to make sure their answer makes sense. Students use division strategies to solve real-world measurement number stories, converting between different units of measurement.

Please keep this Family Letter for reference as your child works through Unit 7.

Vocabulary

Important terms in Unit 7:

line plot A sketch of data in which checkmarks, Xs, stick-on notes, or other marks above a labeled line show the frequency of each value.

mixed number A number that is written using both a *whole number* and a *fraction*. For example, $5\frac{2}{3}$ is a mixed number equal to $5 + \frac{2}{3}$.

multiple of a fraction A product of a fraction and a counting number. For example, $\frac{5}{4}$ is a multiple of $\frac{1}{4}$ because $\frac{5}{4} = 5 * \left(\frac{1}{4}\right)$.

unit fraction A fraction in which the numerator is 1. For example, $\frac{1}{4}, \frac{1}{6},$ and $\frac{1}{10}$ are unit fractions. Fractions can be built from unit fractions. For example, $\frac{3}{4}$ can be built from three $\frac{1}{4}$s.

Do-Anytime Activities

To work with your child on concepts taught in this unit, try these activities:

1. Have your child make a list of shoe sizes from the members of the household and create a line plot from the data. Ask questions like these: *What is the largest shoe size? The smallest? What is the difference between the largest and smallest shoe size?*

2. Ask your child to convert weights of common items into fractions of a pound. For example, a 4-ounce tube of toothpaste $= \frac{1}{4}$ pound.

3. Ask questions like these:
    -   *How long did it take you to get to school?*
    -   *What fraction of an hour is that?*
    -   *If it takes you 3 times as long to get to school tomorrow, how long will it take you?*
    -   *How much time do you spend all week getting to school?*

4. Look at a store advertisement or sale flyer and pose questions about items sold in bulk. For example: *What is the cost of 1 _____? What is the cost if we buy _____ or _____?*

Building Skills through Games

In this unit your child will play the following new game to increase his or her understanding of fraction operations. For detailed instructions, see the *Student Reference Book*.

Fraction Multiplication Top-It See *Student Reference Book*, page 264. Students practice multiplying a whole number by a fraction, and they compare their answer with a partner's.

As You Help Your Child with Homework

As your child brings assignments home, it may be helpful to review the instructions together, clarifying them as necessary. The answers listed below will guide you through the Home Links in Unit 7.

Home Link 7-1

1. Answers vary.
3. Answers vary.
5. 4 pints
7. 2 pints
9. 3 quarts
11. 546
13. 4,430

Home Link 7-2

1. $\frac{7}{4}$, or $1\frac{3}{4}$ cups
3. a. $\frac{3}{6}$, or $\frac{1}{2}$ cup
    b. $\frac{15}{6}$, or $2\frac{3}{6}$, or $2\frac{1}{2}$ cups
5. 3,250
7. 22,104

Home Link 7-3

1. $4 * \frac{1}{5} = \frac{4}{5}; \frac{4}{5}$
3. $5 * \frac{1}{2} = \frac{5}{2}$, or $2\frac{1}{2}; \frac{5}{2}$, or $2\frac{1}{2}$ avocados
5. $\frac{3}{2}$, or $1\frac{1}{2}$
7. $\frac{5}{10}$, or $\frac{1}{2}$

Home Link 7-4

1. $\frac{5}{5}$, or 1
3. $\frac{18}{6}$, or 3
5. $5 * \frac{6}{10} = \frac{30}{10}$, or 3 miles
    $7 * \frac{6}{10} = \frac{42}{10}$, or $4\frac{2}{10}$, or $4\frac{1}{5}$ miles
7. 2,096
9. 14,752

Home Link 7-5

1. $5 * 1\frac{1}{2} = l; \frac{15}{2}$, or $7\frac{1}{2}$ pounds; 7 and 8; 120 ounces

3. $14\frac{3}{6}$; 14 and 15

5. $\frac{6}{4}$, or $1\frac{2}{4}$ **7.** $\frac{3}{6}$

Home Link 7-6

1. $8 * \frac{3}{8} = \frac{24}{8}$, or 3 pounds

3. $4 * \frac{5}{8}$ lb $= \frac{20}{8}$, or $2\frac{4}{8}$ pounds

5. 45 R1   **7.** 192 R3

Home Link 7-7

1. A; \$2 more per ticket; Sample answer:
276 ÷ 2 = 138; 138 ÷ 6 = 23; 336 ÷ 2 = 168;
168 ÷ 8 = 21

3. 4,524   **5.** 5,817

Home Link 7-8

1. Sample answer: $(5 * 1,000) - (8 * 500) = w$; 1,000 milliliters

3. Sample answer: $1,400 - (13 * 100) = p$; 100 centimeters

5. $3\frac{4}{6}$ **7.** $5\frac{2}{12}$

Home Link 7-9

1. The perimeter is 4 times the side length.

3. 125 toothpicks

5. 251   **7.** 31 R4

Home Link 7-10

1. a. Yes; $\frac{5}{2}$, or $2\frac{1}{2}$ miles

  b. $\frac{10}{2}$, or 5 miles; Sample answer: Tony will run $\frac{1}{2}$ mile 5 times a week. $5 * \frac{1}{2} = \frac{5}{2}$ miles. For 2 weeks, add $\frac{5}{2} + \frac{5}{2} = \frac{10}{2}$, or 5 miles. 5 > 4.

3. 321   **5.** 147 R4

Home Link 7-11

1. $\frac{15}{5}$, or 3   **3.** 9

5. $1\frac{2}{4}$ pounds; 24 ounces

7. 116 R2   **9.** 42 R1

Home Link 7-12

1. \$5.53; Sample answer: $7 * \frac{79}{100} = \frac{79}{100} + \frac{79}{100} + \frac{79}{100} + \frac{79}{100} + \frac{79}{100} + \frac{79}{100} + \frac{79}{100} = \frac{553}{100} = 5$ and 53 hundredths $= 5.53$

3. \$1.69; Sample answer: $\frac{1,000}{100} - \frac{831}{100} = \frac{169}{100}$

5. =   **7.** >

Home Link 7-13

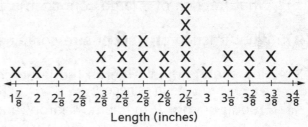

Pencil Lengths (to the nearest $\frac{1}{8}$ inch)

Length (inches)

1. 13 students

3. a. 3 pencils   **b.** 6 inches

5. a. $3\frac{4}{8}$ inches   **b.** $1\frac{7}{8}$ inches

  c. $4\frac{11}{8}$, or $5\frac{3}{8}$ inches   **d.** $1\frac{5}{8}$ inches

7. $12\frac{2}{10}$ **9.** $3\frac{90}{100}$

Shopping for Milk

Solve.

Dale's Dairy sells milk in containers of various sizes as shown below.

Container Size	Amount
Small	1 cup
Medium	1 pint
Large	1 quart
Extra large	$\frac{1}{2}$ gallon
Super size	1 gallon

(1) What fraction of a large container is 1 cup? _____

(2) What fraction of a super size container is 1 pint? _____

(3) How many times larger is the large container than the small? _____ times

(4) Freddy bought 2 medium, 4 large, and 2 extra large containers of milk. He needs $2\frac{1}{2}$ gallons of milk for the week. How much more does he need?

_____

(5) Thomas bought 5 large and 2 medium containers of milk. Now he has 3 gallons of milk all together. How much did he have before he went to Dale's?

_____

(6) Elsa needs $2\frac{1}{2}$ gallons of milk to make pudding for a family reunion. The dairy is out of the three largest-size containers. What amounts should Elsa buy to have enough milk?

_____

(7) Write your own number story about Dale's Dairy and solve it.

_____

_____

_____

Liquid Measures

Find at least one container that holds each of the amounts listed below.
Describe each container and record all the measurements on the label.

SRB
195

① About 1 gallon

Container	Liquid Measurements on Label
jug of orange juice	*gallon, 3.78 L*

② About 1 quart

Container	Liquid Measurements on Label
container of milk	*1 quart, 32 fl oz*

③ About 1 pint

Container	Liquid Measurements on Label

④ About 1 cup

Container	Liquid Measurements on Label

Complete.

⑤ 2 quarts = _____ pints

⑥ 3 gallons = _____ cups

⑦ _____ pints = 4 cups

⑧ _____ quarts = 12 cups

⑨ 6 pints = _____ quarts

⑩ _____ quarts = $2\frac{1}{2}$ gallons

Practice

⑪ 273 * 2 = _____

⑫ 385 * 4 = _____

⑬ _____ = 886 * 5

⑭ _____ = 98 * 38

261

Using Measuring Cups and Spoons

For each problem, measure the given amount of water, sand, or other material in the smaller measuring cup or spoon. Pour the contents into the larger measuring tool. Repeat until the larger measuring tool is full. Write a fraction addition number sentence to show what you found.

Use the measuring cups to complete Problems 1–4:

1. How many times do you have to fill the $\frac{1}{3}$ cup to fill 1 cup? ____

   Number model: $\frac{1}{3} + \frac{1}{3} + \frac{1}{3} = 1$

2. How many times do you have to fill the $\frac{1}{4}$ cup to fill the $\frac{1}{2}$ cup? ____

   Number model: _____ $= \frac{1}{2}$

3. How many times do you have to fill the $\frac{1}{2}$ cup to fill $1\frac{1}{2}$ cups? ____

   Number model: _____ $= 1\frac{1}{2}$

4. How many times do you have to fill the $\frac{1}{4}$ cup to fill $1\frac{1}{2}$ cups? ____

   Number model: _____ $= 1\frac{1}{2}$

Use the measuring spoons to complete Problems 5–7.

5. How many times do you have to fill the $\frac{1}{4}$ teaspoon to fill the $\frac{1}{2}$ teaspoon? ____

   Number model: _____ $= \frac{1}{2}$

6. How many times do you have to fill the $\frac{1}{2}$ teaspoon to fill 2 teaspoons? ____

   Number model: _____ $= 2$

7. How many times do you have to fill the $\frac{1}{2}$ teaspoon to fill $2\frac{1}{2}$ teaspoons? ____

   Number model: _____ $= 2\frac{1}{2}$

262

Sugar in Drinks

Use the information in the table to solve the number stories. In the space below each problem, use pictures or equations to show what you did to find your answers.

Amount of Sugar in Drinks		
Drink	**Sugar Content (in cups)**	**Serving Size (in ounces)**
Cranberry juice cocktail	$\frac{1}{4}$	12
Fruit punch	$\frac{1}{4}$	12
Orange soda	$\frac{1}{4}$	12
Sweet tea	$\frac{1}{6}$	12

Sources: National Institutes of Health and California Department of Public Health

① Carmen drinks one 12-ounce can of orange soda every day. How much sugar is that in 1 week? _____ cup(s)

② If you drink one 12-ounce glass of cranberry juice cocktail every morning, how much sugar will that be in 2 weeks? _____ cup(s)

③ Mike drinks three 12-ounce servings of sweet tea per day.

a. How much sugar is he drinking in his tea in 1 day?

_____ cup(s)

b. In 5 days? _____ cup(s)

Practice

④ 951 * 4 = _____

⑤ 650 * 5 = _____

⑥ 425 * 7 = _____

⑦ 3,684 * 6 = _____

263

Skip Counting by a Unit Fraction

① Use your calculator to count by $\frac{1}{2}$s. Complete the table below.

One $\frac{1}{2}$	Two $\frac{1}{2}$s	Three $\frac{1}{2}$s	Four $\frac{1}{2}$s	Five $\frac{1}{2}$s	Six $\frac{1}{2}$s	Seven $\frac{1}{2}$s	Eight $\frac{1}{2}$s	Nine $\frac{1}{2}$s	Ten $\frac{1}{2}$s
$\frac{1}{2}$	$\frac{2}{2}$	$\frac{3}{2}$	$\frac{4}{2}$						

② Use your calculator to count by $\frac{1}{3}$s. Complete the table below.

One $\frac{1}{3}$	Two $\frac{1}{3}$s	Three $\frac{1}{3}$s	Four $\frac{1}{3}$s	Five $\frac{1}{3}$s	Six $\frac{1}{3}$s	Seven $\frac{1}{3}$s	Eight $\frac{1}{3}$s	Nine $\frac{1}{3}$s	Ten $\frac{1}{3}$s
$\frac{1}{3}$	$\frac{2}{3}$	$\frac{3}{3}$	$\frac{4}{3}$						

③ Use your calculator to count by $\frac{1}{5}$s. Complete the table below.

One $\frac{1}{5}$	Two $\frac{1}{5}$s	Three $\frac{1}{5}$s	Four $\frac{1}{5}$s	Five $\frac{1}{5}$s	Six $\frac{1}{5}$s	Seven $\frac{1}{5}$s	Eight $\frac{1}{5}$s	Nine $\frac{1}{5}$s	Ten $\frac{1}{5}$s
$\frac{1}{5}$	$\frac{2}{5}$								

④ Use your calculator to count by $\frac{1}{8}$s. Complete the table below.

One $\frac{1}{8}$	Two $\frac{1}{8}$s	Three $\frac{1}{8}$s	Four $\frac{1}{8}$s	Five $\frac{1}{8}$s	Six $\frac{1}{8}$s	Seven $\frac{1}{8}$s	Eight $\frac{1}{8}$s	Nine $\frac{1}{8}$s	Ten $\frac{1}{8}$s
$\frac{1}{8}$									

⑤ Use your calculator to count by $\frac{1}{10}$s. Complete the table below.

One $\frac{1}{10}$	Two $\frac{1}{10}$s	Three $\frac{1}{10}$s	Four $\frac{1}{10}$s	Five $\frac{1}{10}$s	Six $\frac{1}{10}$s	Seven $\frac{1}{10}$s	Eight $\frac{1}{10}$s	Nine $\frac{1}{10}$s	Ten $\frac{1}{10}$s

⑥ How is skip counting by $\frac{1}{3}$s on your calculator from 0 to nine $\frac{1}{3}$s the same as finding the product $9 * \frac{1}{3}$?

_____

_____

_____

Finding Multiples of Unit Fractions

For Problems 1–3, fill in the blanks to complete an equation describing the number line.

SRB
171-174

①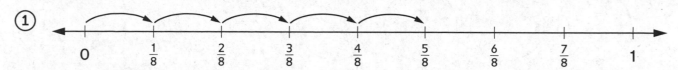

Equation: 5 * _____ = _____

②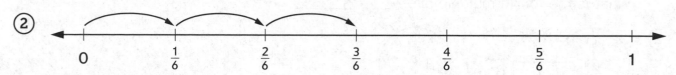

Equation: _____ * $\frac{1}{6}$ = _____

③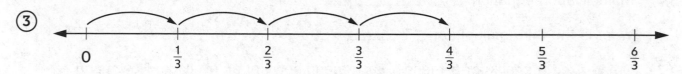

Equation: _____ * _____ = _____

For Problems 4–6, use the number line to help you multiply the fraction by the whole number.

④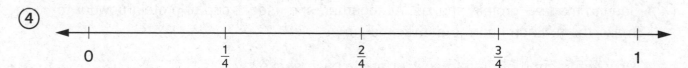

Equation: 2 * $\frac{1}{4}$ = _____

⑤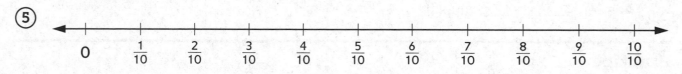

Equation: 6 * $\frac{1}{10}$ = _____

⑥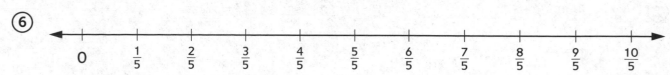

Equation: 7 * $\frac{1}{5}$ = _____

Multiplying Unit Fractions

Write a multiplication equation to describe each picture or story.

SRB
171-174

①

Multiplication equation: _____

What is the fourth multiple of $\frac{1}{5}$? _____

②

Multiplication equation: _____

What is the third multiple of $\frac{1}{10}$? _____

③ Dmitri fixed a snack for 5 friends. Each friend got $\frac{1}{2}$ of an avocado. How many avocados did Dmitri use?

Multiplication equation: _____

Answer: _____ avocado(s)

④ Juanita made 3 protein shakes. All together, she used 1 cup of protein powder to make them. Each had the same amount.

How many cups of protein powder are in each shake?

Multiplication equation: _____

Answer: _____ cup(s)

Practice

⑤ $\frac{1}{2} + \frac{1}{2} + \frac{1}{2} =$ _____

⑥ $\frac{2}{3} + \frac{2}{3} + \frac{1}{3} =$ _____

⑦ $\frac{9}{10} - \frac{4}{10} =$ _____

⑧ $\frac{8}{12} - \frac{5}{12} =$ _____

Multiplying Fractions Using an Addition Model

Draw models for each product. Then add the fractions to find the product.

SRB
173-174

① $2 * \frac{1}{3} =$ _____

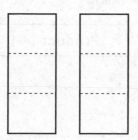

② $3 * \frac{1}{2} =$ _____

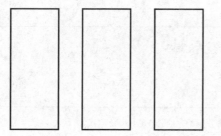

③ $2 * \frac{2}{5} =$ _____

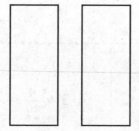

④ $4 * \frac{2}{3} =$ _____

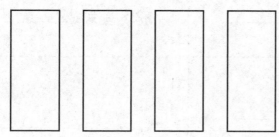

Products < or > 1

Find and record the product for each equation at the bottom of the page. Create one of your own in the empty box. Then cut out each equation. Glue the equation in either the Product < 1 or Product > 1 column in the table below.

Product < 1	Product > 1

✂

$2 * \frac{1}{4} =$ _____

$5 * \frac{1}{10} =$ _____

$3 * \frac{4}{10} =$ _____

$4 * \frac{2}{3} =$ _____

$7 * \frac{4}{10} =$ _____

$2 * \frac{3}{5} =$ _____

$3 * \frac{1}{10} =$ _____

$3 * \frac{2}{8} =$ _____

Multiplying Fractions by Whole Numbers

Solve the problems below.

① $5 * \frac{1}{5} =$ _____

Draw a picture.

② $3 * \frac{4}{9} =$ _____

Draw a picture.

③ $6 * \frac{3}{6} =$ _____

Draw a picture.

Write a multiplication equation to represent the problem and then solve.

④ Rahsaan needs to make 5 batches of granola bars. A batch calls for $\frac{1}{2}$ cup of honey.

How much honey does he need? Equation: _____

⑤ Joe swims $\frac{6}{10}$ of a mile 5 days per week. How far does he swim every week?

Equation: _____

How far would he swim if he swam every day of the week?

Equation: _____

Practice

⑥ $653 * 3 =$ _____

⑦ $262 * 8 =$ _____

⑧ $357 * 9 =$ _____

⑨ $7,376 * 2 =$ _____

269

Multiple Solutions

The table below shows two different ways to solve each problem. Complete both solutions to find the answer to each problem.

Problem	Solution 1	Solution 2
$6 * 3\frac{2}{3}$	$6 * \frac{11}{3} =$	$6 * \left(3 + \frac{2}{3}\right) =$
$3 * 4\frac{5}{8}$	$3 * \left(4 + \frac{5}{8}\right) =$	$3 * \frac{37}{8} =$
$5 * 1\frac{4}{5}$	$5 * \left(1 + \frac{4}{5}\right) =$	$(5 * 1) + \left(5 * \frac{4}{5}\right) =$
$2 * 5\frac{3}{10}$	$(2 * 5) + \left(2 * \frac{3}{10}\right) =$	$2 * \frac{53}{10} =$
$4 * 2\frac{5}{6}$	$4 * \left(2 + \frac{5}{6}\right) =$	$(4 * 2) + \left(4 * \frac{5}{6}\right) =$

Multiplying Mixed Numbers by Whole Numbers

Solve.

1. Michelle's grandmother sent her 5 small gifts for her fifth birthday. Each one weighed $1\frac{1}{2}$ pounds. How much did the gifts weigh all together?

   Number model with unknown: _____

   Answer: _____ pounds

   Between what two whole numbers is this? _____ and _____

   How many ounces did the gifts weigh? _____ ounces

2. Rochelle bought 4 pieces of ribbon to finish a project. Each piece was $1\frac{5}{12}$ yards long. What is the combined length of the ribbon she bought?

   Number model with unknown: _____

   Answer: _____ yards

   Between what two whole numbers is this? _____ and _____

   How many feet is this? _____ feet

3. $3 * 4\frac{5}{6} =$ _____

   Between what two whole numbers is this? _____ and _____

4. $6 * 7\frac{3}{8} =$ _____

   Between what two whole numbers is this? _____ and _____

Practice

5. $\frac{3}{4} + \frac{2}{4} + \frac{1}{4} =$ _____

6. $\frac{4}{8} + \frac{3}{8} + \frac{2}{8} =$ _____

7. $\frac{5}{6} - \frac{2}{6} =$ _____

8. $\frac{88}{100} - \frac{57}{100} =$ _____

271

Three-Fruit Salad

The school cook asks you to create recipes for Three-Fruit Salad. Follow these rules:

- Each recipe must use exactly 3 different fruits.

- The combined weight of the fruit for one recipe must be exactly 5 pounds.

Fruit	Typical Weight
Medium pear	$\frac{3}{8}$ lb
Cup of grapes	$\frac{2}{8}$ lb
Large orange	$\frac{6}{8}$ lb
Pint of strawberries	$\frac{5}{8}$ lb
Medium apple	$\frac{4}{8}$ lb

Make up two recipes that follow the rules. Show that each recipe weighs
5 pounds by using tools such as fraction circles, fraction number lines, drawings,
or number models. Use multiplication when possible.

Recipe #1

Three-Fruit Salad (continued)

Recipe #2

Fruit Salad Weight

Mr. Chou makes fruit salad that he sells in his store. Today he plans to make a fruit salad with 8 pears, 2 cups of grapes, and 4 pints of strawberries. Use the weights below to solve the problems.

- A medium pear weighs about $\frac{3}{8}$ lb.

- A cup of grapes weighs about $\frac{2}{8}$ lb.

- A pint of strawberries weighs about $\frac{5}{8}$ lb.

(1) Write a multiplication sentence to show how much the pears weigh. _____

Answer: _____ pound(s)

(2) Write a multiplication sentence to show how much the grapes weigh. _____

Answer: _____ pound(s)

(3) Write a multiplication sentence to show how much the strawberries weigh.

_____

Answer: _____ pound(s)

(4) How much does Mr. Chou's salad weigh in all? Show your work.

Answer: _____ pound(s)

Practice

(5) 361 / 8 = _____

(6) 396 ÷ 7 = _____

(7) 963 / 5 = _____

(8) 633 / 4 = _____

Solving Multistep Number Stories

For each problem, estimate the answer and then solve. Write number models with letters for the unknowns to show what you did.

SRB
111-115

① The River Forest Pet Shelter had 173 dogs and 149 cats. On Tuesday the shelter lost power during a storm. Nearby shelters agreed to take animals until the power was restored. Oak Shelter took 49 dogs and 55 cats. The Wyn and May shelters said they would split the rest of the dogs and cats equally. How many dogs did Wyn and May take? How many cats?

Estimate: _____

The Wyn and May shelters each took _____ dogs and _____ cats.

Number model(s) with unknown(s): _____

② On Monday Wyn Pet Shelter had 128 male dogs and 26 females. On Tuesday they took in the dogs from the River Forest shelter and found homes for 16 dogs. If volunteers walk the dogs they have left at the end of the day in groups of 8, how many groups will they have to walk?

Estimate: _____

The volunteers walked _____ groups of dogs.

Number model(s) with unknown(s): _____

③ The May Pet Shelter began the day Tuesday with 75 male cats and 32 females. They took in the cats from the River Forest shelter and found homes for 13 cats during the day. If they split the cats they have left at the end of the day into groups of 3 to be fed, how many groups will they have?

Estimate: _____

The May shelter will have _____ groups of cats to feed.

Number model(s) with unknown(s): _____

Division Number Stories

Solve. Show your work.

SRB
111-115

(1) Robert and Jason want to buy a group ticket package for football games. Package A costs $276 and includes 2 tickets for each of 6 games. Package B costs $336 and includes 2 tickets for each of 8 games. Which package charges more per ticket? How much more per ticket?

Package _____ charges $ _____ more per ticket.

(2) Rebecca wants to put 544 pennies in a coin-collection book. The blue book fits 9 pennies per page. The red book fits 7 pennies per page. How many more pages would she need if she used the red book rather than the blue one?

The red book will take _____ more pages than the blue book.

What did you do with any remainders you found?

_____

_____

_____

Practice

(3) 754 * 6 = _____

(4) 906 * 2 = _____

(5) _____ = 831 * 7

(6) _____ = 84 * 29

276

Soccer Field Measurements

Use the information below to write a number model with an unknown and then solve.

Field Dimensions and Match Times for Outdoor and Indoor Soccer		
	Outdoor	Indoor
Length of field	100 meters	60 meters
Width of field	75 meters	25 meters
Distance between goal posts	8 yards	14 feet
Time of match	90 minutes	60 minutes

① Gregg wants to warm up for his outdoor soccer game. He runs 400 meters.

   a. How many lengths of the field did he run?

   Number model with unknown: _____   Answer: _____ lengths

   b. How many centimeters did he run?   Answer: _____ centimeters

② Indoor soccer has 4 periods per match.

   a. How long is each period?

   Number model with unknown: _____   Answer: _____ minutes

   b. How long is this in seconds? _____ seconds

③ Outdoor soccer has 2 halves per match.

   a. How long is each half?

   Number model with unknown: _____   Answer: _____ minutes

   b. How long is this in seconds? _____ seconds

④ Parents are putting in the goalposts for the new outdoor soccer fields.

   a. What is the halfway point between the two posts?

   Number model with unknown: _____   Answer: _____ yards

   b. How many feet is this? _____ feet

   c. What is the difference in feet between the length of half of an outdoor soccer goal and the length of a full indoor soccer goal? _____ feet

More Division Measurement Number Stories

Read each number story. Use the information to write a number model with an unknown and then solves.

(1) Kelly is in charge of bringing water for her softball game. The 8 members of the team have matching team water bottles that hold 500 mL. Kelly buys 5 liters of water at the store. If she fills all the bottles, how many milliliters of water will Kelly have left?

Number model with unknown: _____

Answer: _____ milliliters

(2) The distance around all the bases in softball is 72 meters. If Kelly hits 2 home runs and runs around the bases twice, how many millimeters will she run?

Number model with unknown: _____

Answer: _____ millimeters

(3) In women's softball the pitcher stands about 13 meters from the batter's box. In men's softball the pitcher stands about 1,400 centimeters from the batter's box. About how many more centimeters is it from the men's pitcher to the batter's box than from the women's pitcher to the batter's box?

Number model with unknown: _____

Answer: About _____ centimeters

(4) The 6 games Kelly's team played took a total of 7 hours.

a. How many minutes total did they play softball?

Number model with unknown: _____

Answer: _____ minutes

b. If each game lasted the same amount of time, how many minutes did each one last?

Number model with unknown: _____

Answer: _____ minutes

Practice

(5) $1\frac{3}{6} + 2\frac{1}{6} =$ _____

(6) $4\frac{3}{5} + 5\frac{4}{5} =$ _____

(7) $7\frac{5}{12} - 2\frac{3}{12} =$ _____

(8) $6\frac{1}{3} - 2\frac{2}{3} =$ _____

Finding Patterns

① Fill in the patterns and the rule:

a. 2, 4, 6, _____, _____, _____   Rule: _____

b. 5, 10, 15, _____, _____, _____   Rule: _____

c. 15, 12, 9, _____, _____, _____   Rule: _____

d. 90, 80, 70, _____, _____, _____   Rule: _____

② **a.** Draw a square array with 4 dots.   **b.** Draw a square array with 9 dots.

c. Draw a square array with 16 dots.   **d.** What should the next square array look like?

e. Explain how you know what the array for Problem 2d should look like.

_____

_____

_____

_____

Trading Cards

Draw pictures or make diagrams or charts to help you answer the questions below.

(1) **a.** If Mary and Tran each gave each other 1 card, how many cards were traded in all?

_____ cards

b. If Mary, Tran, and Helga each had 2 cards, and all gave each other a card, how many cards were traded in all?

_____ cards

c. If Mary, Tran, Helga, and Philippe each had 3 cards, and all gave each other a card, how many cards were traded in all?

_____ cards

(2) **a.** Fill in the first two lines of the chart using your answers from above. Find the pattern and use it to fill in the rest of the chart. Then answer the questions below.

People	Cards	Equation
2		$2 * (2 - 1) = 2$
3		$3 * (3 - 1) = 6$
4		
5	20	$5 * (5 - 1) = 20$
6	30	
7		

b. What rule describes the pattern in the chart? _____

_____

_____

c. Describe other patterns you notice in the chart. _____

_____

_____

_____

280

Perimeter Patterns

Alice was making squares out of toothpicks. She noticed a pattern involving the length of one side and the perimeter of the square. Complete the table and then answer the questions that follow.

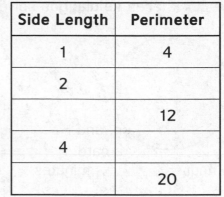

Side Length	Perimeter
1	4
2	
	12
4	
	20

SRB
200-201, 58-59

① What rule describes the relationship between the length of one side and the perimeter of a square?

_____

② What would be the perimeter of a square with a side length of 25 toothpicks?

_____ toothpicks

③ What would be the side length of a square with a perimeter of 500 toothpicks?

_____ toothpicks

④ Describe at least two other patterns you notice in the table _____

_____

_____

_____

_____

_____

_____

_____

Practice

⑤ $753 \div 3 =$ _____

⑥ _____ $= 386 \div 2$

⑦ $283 \div 9 \rightarrow$ _____

⑧ $505 \div 6 \rightarrow$ _____

Fractions of Hours

Use the clock faces to help you solve the problems.

(1)

$\frac{1}{2}$ hour = _____ minutes

$\frac{2}{2}$ hour = 2 * 30 = _____ minutes

(2)

$\frac{1}{6}$ hour = 10 minutes

$\frac{2}{6}$ hour = 2 * 10 = _____ minutes

$\frac{4}{6}$ hour = 4 * 10 = _____ minutes

(3)

$\frac{1}{12}$ hour = 5 minutes

$\frac{5}{12}$ hour = _____ * _____ = _____ minutes

$\frac{7}{12}$ hour = _____ * _____ = _____ minutes

(4)

$\frac{1}{10}$ hour = 6 minutes

$\frac{3}{10}$ of 60 = _____ * _____ = _____ minutes

$\frac{9}{10}$ of 60 = _____ * _____ = _____ minutes

(5)

$\frac{1}{3}$ hour = 20 minutes

$\frac{2}{3}$ hour = _____ * _____ = _____ minutes

$\frac{3}{3}$ hour = _____ * _____ = _____ minutes

(6)

$\frac{1}{4}$ hour = 15 minutes

$\frac{2}{4}$ hour = _____ * _____ = _____ minutes

$\frac{3}{4}$ hour = _____ * _____ = _____ minutes

Solving Multistep Number Stories with Fractions

① Josiah is painting his bedroom. Each of his 4 walls will need $\frac{3}{4}$ gallon of paint. He has $4\frac{2}{4}$ gallons of paint. How much will he have left over?

Number model with unknown: _____

Answer: _____ gallon(s)

How many quarts is this? _____ quart(s)

② Dartrianna's bedroom is slightly larger than Josiah's. Each of her 4 walls will need $\frac{7}{8}$ gallon of paint. She has $3\frac{3}{8}$ gallons of paint. How much more does she need?

Number model with unknown: _____

Answer: _____ gallon(s)

How many pints is this? _____ pint(s)

③ Josiah thinks it will take him $\frac{1}{3}$ hour to paint each wall. Dartrianna thinks it will take her $\frac{2}{3}$ hour to paint each wall. How much longer will it take Dartrianna to paint her room than it will take Josiah?

Number model with unknown: _____

Answer: _____ hours

How many minutes is this? _____ minute(s)

④ Omaria painted the garage. It took her $2\frac{4}{10}$ hours to paint each of the 2 larger walls and $1\frac{7}{10}$ hours to paint each of the 2 smaller walls. How much time did it take her to paint all four walls?

Number model with unknown: _____

Answer: _____ hour(s)

How many minutes is this? _____ minute(s)

Fitness Challenge

Use the information in the table below to solve the number stories.

During Marcy School's 2-week challenge, each student who meets a goal wins a prize.

Marcy's Fitness Challenge Goals

Activity	Total Distance	Activity	Total Distance
Walking	6 miles	Bike Riding	6 miles
Swimming	1 mile	Running	4 miles

(1) Tony will run $\frac{1}{2}$ mile after school each day. Will he win a prize? _____

    a. Distance run in 1 week:_____ mile(s)   **b.** In 2 weeks: _____ mile(s)

    Explain how you found your answer.

_____

_____

(2) Three times a week, Tina walks $\frac{3}{10}$ mile from school to the library, studies for 1 hour, and then walks $\frac{4}{10}$ mile home. How much more will she need to walk to win a prize?

    _____ mile(s)

    Explain how you found your answer.

_____

_____

_____

Practice

(3) $642 \div 2 =$ _____

(4) $386 / 9 \rightarrow$ _____

(5) $739 / 5 \rightarrow$ _____

(6) $4\overline{)829} \rightarrow$ _____

Converting Pounds to Ounces

Pounds	Ounces	Object in Our Room
1	16	
2		
3		
5		
$\frac{7}{8}$		
$\frac{6}{8}$ $\left(\text{or } \frac{3}{4}\right)$		
	10	
$\frac{4}{8}$ $\left(\text{or } \frac{1}{2}\right)$		
$\frac{3}{8}$		
$\frac{2}{8}$ $\left(\text{or } \frac{1}{4}\right)$		
$\frac{1}{8}$	2	

Fractions and Mixed Numbers

Solve. Draw a picture or show how you solved the problem.

① $5 * \frac{3}{5} =$ _____

② _____ $= 4\frac{2}{6} - 2\frac{4}{6}$

③ $5\frac{7}{8} + 3\frac{1}{8} =$ _____

④ _____ $= 3 * 4\frac{1}{4}$

⑤ The combined weight of an assortment of fruit is $8\frac{3}{4}$ pounds. When the fruit is on a tray, the tray weighs $10\frac{1}{4}$ pounds. How many pounds does the tray weigh when empty? _____ pound(s)

How many ounces does the tray weigh when empty? _____ ounce(s)

⑥ $\left(3 * 2\frac{2}{3}\right) + \left(2 * 4\frac{1}{3}\right) =$ _____

Practice

⑦ $3\overline{)350}$ ⑧ $6\overline{)832}$

⑨ $7\overline{)295}$ ⑩ $9\overline{)582}$

Finding Fraction and Decimal Equivalents

Represent each fraction and decimal in different ways.

Fraction	Decimal	Number Name	Money (dollars, dimes, pennies)
		four-tenths	
		thirty-six hundredths	3 dimes, 6 pennies
$\frac{7}{100}$			
	4.52		

Making Goodie Bags

Help create goodie bags for 4 different birthday parties. Each birthday boy or girl wants a different number of items and has a different budget per goodie bag.

For each party's goodie bag, name each item, write how many of that item will be in one bag, and write the total cost for this item per bag. Then find the total cost for all of the items in each party's goodie bag.

Item	Cost
Erasers	$0.16 each
Key chains	$0.59 each
Rubber balls	$0.20 each
Markers	$1.40 each
Stickers	$1.39 per pack
Tattoos	$0.10 each
Small baseballs	$0.49 each
Bookmarks	$0.78 each
Crayons	$0.99 per box
Marbles	$1.41 per bag
Whistles	$0.18 each

Example: Exactly 4 items per bag. Total cost per bag: between $1.00 and $1.50.

1 bookmark: $0.78, or $\frac{78}{100}$; 3 tattoos: $3 * \frac{10}{100} = \frac{30}{100}$, or $0.30; Total: $\frac{108}{100}$ or $1.08

① For Leonard's party: Exactly 5 items per bag. Total cost per bag: between $1.50 and $1.75.

_____

_____

_____

② For Simeon's party: Exactly 5 different items per bag. Total cost per bag: between $3.25 and $3.50.

_____

_____

_____

③ For Madge and Mitch's party: Exactly 8 different items per bag. Total cost per bag: between $4.50 and $5.00.

_____

_____

_____

④ Write a description for another goodie bag that allows someone to use more than one of the same item. Trade with a partner to solve.

Shopping for Bargains

Solve each number story and show how you solved the problems.

SRB
166-167,
175-176

(1) Phil wants to buy some Creepy Creature erasers that cost $1.05 each.
If he buys 5 or more, the price is $0.79 each. If he decides to buy 7 erasers,
how much will he spend?

Answer: $_____

(2) Mrs. Katz bought 3 pounds of apples and a muffin for snacks. The apples cost $2.59
per pound if you buy less than 3 pounds and $2.12 per pound if you buy 3 or more
pounds. The muffin cost $1.95. How much did she spend?

Answer: $_____

Try This

(3) Mrs. Katz paid with a $10 bill. How much change did she get back?

Answer: $_____

Practice

Fill in the blanks with >, <, or =.

(4) 0.55 ____ 0.65   (5) 0.3 ____ 0.30   (6) 0.72 ____ 0.8   (7) 0.4 ____ 0.31

Math Message

1. Lyric, Chloe, and Matthew measured line segment AB to the nearest $\frac{1}{8}$ inch. Lyric said it was $2\frac{4}{8}$ inches long, Chloe said it was $2\frac{2}{8}$ inches long, and Matthew said it was $2\frac{1}{2}$ inches long. With whom do you agree? _____
Why?

_____

_____

A _____ B

2. Measure line segment CD to the nearest $\frac{1}{8}$ inch. _____ inches

C _____ D

✂ --

Math Message

Lesson 7-13

NAME   DATE   TIME

1. Lyric, Chloe, and Matthew measured line segment AB to the nearest $\frac{1}{8}$ inch. Lyric said it was $2\frac{4}{8}$ inches long, Chloe said it was $2\frac{2}{8}$ inches long, and Matthew said it was $2\frac{1}{2}$ inches long. With whom do you agree with? _____
Why?

_____

_____

A _____ B

2. Measure line segment CD to the nearest $\frac{1}{8}$ inch. _____ inches

C _____ D

Mystery Line Plot

① Examine the line plot below. Think about what kind of data it could be used to represent.

Fill in the title and label the horizontal axis.

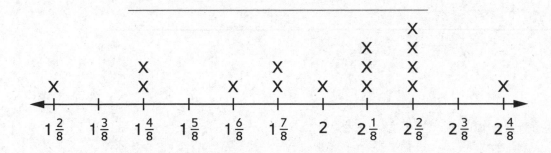

② Write three problems using the line plot above. Show how to solve each problem in the space provided.

a. Problem: _____

Answer: _____

b. Problem: _____

Answer: _____

c. Problem: _____

Answer: _____

Dog Walking Distances

Corey has a dog walking business. She kept track of how many miles she walked each day for 2 weeks:

$\frac{3}{8}$, $\frac{4}{8}$, 1, $\frac{4}{8}$, $\frac{6}{8}$, $\frac{3}{8}$, $\frac{5}{8}$, $\frac{4}{8}$, $\frac{3}{8}$, $\frac{5}{8}$, 1, $\frac{5}{8}$, $\frac{1}{8}$, $\frac{5}{8}$

(1) Plot Corey's data on the line plot below.

Title: _____

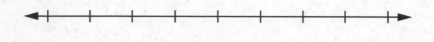

_____

(2) What is the farthest she walked in 1 day? _____ mile(s)

(3) What is the shortest distance she walked in 1 day? _____ mile(s)

(4) How many times did she walk $\frac{2}{8}$ mile? _____ time(s)

(5) What is the distance that she walked most often? _____ mile(s)

(6) How far did she walk all together on days that she walked less than $\frac{1}{2}$ mile?

_____ mile(s)

(7) How many times did she walk $\frac{5}{8}$ mile? _____ time(s)

(8) What is her total distance on the days she walked $\frac{5}{8}$ mile? _____ mile(s)

Pencil Lengths

At the beginning of the year Mrs. Kerry gave each student in her class a new pencil with "Welcome to 4th Grade" written on it. A month later the class measured their pencils to the nearest $\frac{1}{8}$ inch.

Pencil Lengths to the Nearest $\frac{1}{8}$ inch

$2\frac{1}{8}$	$3\frac{1}{8}$	$2\frac{7}{8}$	$2\frac{4}{8}$	$3\frac{3}{8}$	$2\frac{7}{8}$	3	$2\frac{5}{8}$	$2\frac{5}{8}$	$2\frac{7}{8}$	$3\frac{3}{8}$	$2\frac{6}{8}$	$2\frac{4}{8}$
$2\frac{3}{8}$	$2\frac{7}{8}$	$1\frac{7}{8}$	$3\frac{2}{8}$	$2\frac{7}{8}$	$3\frac{4}{8}$	$2\frac{6}{8}$	$2\frac{3}{8}$	$3\frac{1}{8}$	2	$2\frac{4}{8}$	$2\frac{5}{8}$	$3\frac{2}{8}$

Plot the data set on the line plot.

Title: _____

_____

Pencil Lengths

(continued)

Use the completed line plot to answer these questions.

① How many students have a pencil that is shorter than $2\frac{7}{8}$ inches?

_____ students

② What is the most common pencil length? _____ inches

③ a. How many pencils are less than $2\frac{2}{8}$ inches long? _____ pencils

b. What is their combined length? _____ inches

④ a. How many pencils are between $2\frac{7}{8}$ and $3\frac{2}{8}$ inches long? _____ pencils

b. What is their combined length? _____ inches

⑤ a. How long is the longest pencil? _____ inches

b. How long is the shortest pencil? _____ inches

c. What is the combined length of the longest and shortest pencils? _____ inches

d. What is the difference in length of the longest and shortest pencils?

_____ inches

Practice

⑥ $2\frac{1}{4} + 5\frac{2}{4} =$ _____

⑦ $8\frac{5}{10} + 3\frac{7}{10} =$ _____

⑧ $3\frac{7}{8} - 1\frac{3}{8} =$ _____

⑨ $7\frac{41}{100} - 3\frac{51}{100} =$ _____

294

Unit 8: Family Letter

Fraction Operations; Applications

Think back to how you learned to ride a bike as a child. What if you were allowed to practice only on a stationary bike rather than a real one? When you finally ventured out onto the neighborhood streets expecting to ride like a pro, you would probably be disappointed! Without an opportunity to apply what you learned to a real-world situation, you would never have to apply the brakes going down a hill or maneuver around a sharp curve. Likewise, if students aren't given a chance to apply what they learn in mathematics to real-world situations, it may seem to them like useless knowledge. To help make mathematics more meaningful to students, Unit 8 asks them to apply what they have learned throughout the year to real-world problems.

Fraction Operations

This year students have explored adding, subtracting, and multiplying fractions. In Unit 8 they will apply fraction and mixed-number operations to help them solve real-world problems involving the perimeter and area of rectangles and units of measure. For example, students will use the relationship between perimeter and area to find the missing side length of a fence or determine the fractional amounts of juice needed to make fruit punch.

Angle Applications

Angles play important roles in many real-life situations, including carpentry, measuring the angles of the sun, and many sports. Lesson 8-2 uses hockey to demonstrate real-world applications of students' knowledge of angles. For instance, when a hockey player wants to pass the puck and an opponent is blocking the path, the passer hits the puck off the boards at an angle, causing the puck to travel around the opponent. This is called "banking the puck." In Lesson 8-2 students also use what they have learned about angles to explore the role angles play in our field of vision, which is the angle that includes the area that can be seen without moving the head or eyes.

More Applications

In Lesson 8-4 students apply their knowledge of symmetry to quilting patterns and then create their own quilt based on specified numbers of lines of symmetry. In Lesson 8-5 students use real-world data about envelope sizes from the U.S. Postal Service to create line plots. They then answer questions about the data by adding and subtracting fractions. In Lesson 8-12 students use their knowledge of place value, addition, and subtraction to solve challenging puzzles called cryptarithms. In Lesson 8-13 students find equivalent names for numbers.

Please keep this Family Letter for reference as your child works through Unit 8.

Vocabulary

Important terms in Unit 8:

equivalent names Different ways of naming the same number. For example, $2 + 6$, $4 + 4$, $12 - 4$, $18 - 10$, $100 - 92$, $5 + 1 + 2$, eight, VIII, and ~~HH~~ /// are all equivalent names for 8.

fluid ounce (fl oz) A U.S. customary unit of volume or capacity equal to $\frac{1}{16}$ of a pint, or about 29.6 milliliters.

Do-Anytime Activities

To work with your child on concepts taught in this unit, try these activities:

1. Have your child complete number puzzles found in newspapers, magazines, or online. Discuss with your child how he or she found the solutions.

2. Ask your child to measure a rectangular object such as an envelope, notebook, or room in your home. Have him or her find both the perimeter and the area of the object and then compose a word problem about the measurements.

3. Ask your child to point out items that he or she believes are symmetrical. How many lines of symmetry are there in those items?

4. Have your child point out angles in your home and estimate their measures. Ask your child to add angles together or find missing angles based on these estimates.

5. Show your child a food or beverage container and have him or her locate the liquid volume and convert it to a smaller unit. For instance, a juice box might hold 1 cup of juice, which means it holds 8 fluid ounces of juice.

Building Skills through Games

In Unit 8 students play the following game to increase their understanding of numbers and the properties of numbers. For detailed instructions, see the *Student Reference Book*.

Name That Number See *Student Reference Book*, page 268. This game provides practice representing numbers in different ways, using any or all of the four operations: addition, subtraction, multiplication, and division.

As You Help Your Child with Homework

As your child brings assignments home, you may want to go over instructions together, clarifying them as necessary. The answers listed below will guide you through the Home Links in Unit 8.

Home Link 8-1

1. Team B's car; 27 cm **3.** 180 cm

5. 2,833 R1 **7.** 715 R3

Home Link 8-2

1. 165°; $82° + 83° = f$ **3.** 87°; $3° + w = 90°$

5. 137°; $180° − 43° = s$ **7.** $\frac{5}{3}$, or $1\frac{2}{3}$

9. $\frac{11}{5}$, or $2\frac{1}{5}$

Home Link 8-3

1. 60°; Sample answers: $30° + 30° = 60°$; The measure of each small white rhombus angle is 30°, so two of them make 60°.

3. 16,764 **5.** 4,888

Home Link 8-4

1.

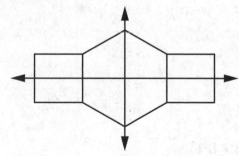

3.

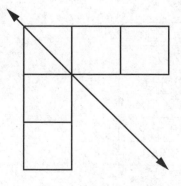

5. 1 line of symmetry;

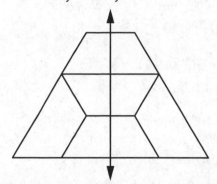

7. $\frac{30}{6}$, or 5 **9.** $\frac{28}{10}$, or $2\frac{8}{10}$

Home Link 8-5

Book Heights

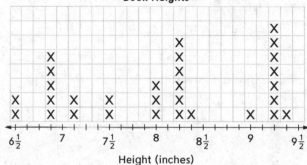

1. $2\frac{7}{8}$ in. **3.** 1,172 R3

Home Link 8-6

1. $7\frac{2}{6}$ yd **3.** Width = $\frac{12}{100}$ km

5. Width = $2\frac{3}{10}$ cm **7.** $\frac{4}{3}$, or $1\frac{1}{3}$

9. $\frac{36}{5}$, or $7\frac{1}{5}$

Home Link 8-7

1. 3.26 kilograms; Sample answer: I thought about what number added to 4 would give me $7\frac{26}{100}$. First I added 3 to get 7. Then I added $\frac{26}{100}$ to get $7\frac{26}{100}$. Finally, $3 + \frac{26}{100} = 3\frac{26}{100} = 3.26$

3. 7.8 cm; Sample answer: $11.4 = 11\frac{4}{10}$ and $3.6 = 3\frac{6}{10}$; $11\frac{4}{10} = 10 + \frac{10}{10} + \frac{4}{10} = 10\frac{14}{10}$; $10\frac{14}{10} - 3\frac{6}{10} = 7\frac{8}{10} = 7.8$

5. 14,316   7. 2,016

Home Link 8-8

1. **a.** $4\frac{1}{12}$ square feet   **b.** $12\frac{8}{12}$ feet

2. **a.** $5\frac{6}{12}$ square feet   **b.** $6\frac{4}{12}$ feet

3. $8\frac{4}{10}$ square inches

5. $\frac{4}{6}$ 7. $\frac{4}{10}$

Home Link 8-9

1. $5\frac{1}{4}$ feet; Sample answer: $3 * 1\frac{3}{4} = (3 * 1) + (3 * \frac{3}{4}) = 3 + \frac{9}{4} = 3\frac{9}{4}$, or $5\frac{1}{4}$

2. Yes. Sample answer: $(5 * 1\frac{1}{2}) + (4 * 1\frac{3}{4}) = 5\frac{5}{2} + 4\frac{12}{4} = 7\frac{1}{2} + 7 = 14\frac{1}{2}$

3. $\frac{6}{6}$, or 1   5. $\frac{54}{100}$

Home Link 8-10

1. Rule: $* 8$

in (gallons)	out (pints)
2	16
$3\frac{1}{2}$	28
6	48
$7\frac{1}{4}$	58
10	80

3. **a.** Yes. Sample answer: The total amount of all the ingredients combined is 18 fluid ounces, so the smoothie will fit in the 24-fluid ounce glass.

b. $\frac{3}{4}$ cup

c. $2\frac{1}{4}$ cups orange juice; 12 fluid ounces cold water; 3 cups vanilla ice cream

d. 54 fluid ounces

5. 1,859   7. 519

Home Link 8-11

1. **a.** $3\frac{1}{8}$ pounds; Sample answer:
$\left(1\frac{1}{2} + \frac{1}{2}\right) + \left(\frac{3}{4} + \frac{1}{4}\right) + \frac{1}{8} = 2 + 1 + \frac{1}{8} = 3\frac{1}{8}$

b. 50 ounces; Sample answer: One pound equals 16 ounces; $\frac{1}{8}$ of a pound = 2 ounces; so $(3 * 16) + 2 = 48 + 2 = 50$

c. 2 packages; Sample answer: 1 of each size uses 50 ounces, so 2 of each size would use $2 * 50 = 100$ ounces. $100 > 80$, so 1 package isn't enough.

2. $1\frac{2}{8}$, or $1\frac{1}{4}$ pounds; Sample answer:
$\left(\frac{1}{8} + \frac{1}{8}\right) + \left(\frac{1}{4} + \frac{3}{4}\right) = \frac{2}{8} + 1 = 1\frac{2}{8}$, or $1\frac{1}{4}$

3. 15,321   5. 2,146

Home Link 8-12

1. Sample answer: $973 + 51 = 1,024$

3. $80 * 64 = 5,120$

5. **a.** 27; $9 * 3 = 27$ **b.** $\frac{1}{3}$; $3 / 9 = \frac{1}{3}$

7. $4\frac{10}{8}$, or $5\frac{2}{8}$ 9. $10\frac{181}{100}$, or $11\frac{81}{100}$

Home Link 8-13

1. Sample answers:

9,990
$2,016 + 7,974$
$(1,427 * 7) + 1$
$1,665 * 6$
$9,000 + 900 + 90$
$13,558 - 3,568$

3. Answers vary.

5. $3\frac{2}{4}$ 7. $2\frac{8}{12}$

Comparing Zoo Admission Costs

Use the information in the chart to solve the number stories below. Prices are rounded to the nearest dollar.

Zoo	Adult (1-Time Admission)	Child (1-Time Admission)	Family Membership (Unlimited Visits)
Los Angeles, CA	$18 each	$13 each	$119 for 2 adults and all children
San Antonio, TX	$12 each	$10 each	$115 for 2 adults and 4 children
Tampa, FL (Lowry Park)	$25 each	$20 each	$160 for 2 adults and 2 children

① How much will 2 adults and 1 child save by buying a family membership instead of paying daily if they go to the Lowry Park Zoo in Tampa 5 times?

Answer: $_____

Number model with unknown: _____

Estimate: _____

② How much will 2 adults and 3 children save by buying the family membership instead of paying daily if they go to the Los Angeles Zoo 3 times?

Answer: $_____

Number model with unknown: _____

Estimate: _____

Does your answer make sense?

_____

③ Five adults took their children to the San Antonio Zoo last Sunday. All together they paid $180. How many children went?

Answer: _____ children

Number model with unknown: _____

Estimate: _____

Does your answer make sense? _____

_____

Multistep Number Stories

The fourth-grade students in Mr. Kennedy's class are investigating energy and motion. Students worked in teams to build two machines: a car that is propelled by a mousetrap and a boat that is propelled by balloons. Today the teams are competing to see which cars and boats go farthest.

Each car or boat gets 3 trials. The total distance from all 3 trials is used to determine which car or boat went farthest. Solve the number stories to help Mr. Kennedy's class compare the machines made by various teams.

1. Team A's car went 173 cm on the first trial, 206 cm on the second trial, and 245 cm on the third trial. Team B's car went 217 cm on each of the three trials.

   Which car went the farthest overall? _____

   How much farther did it go? _____

2. Team A's boat went 130 cm in all. Team B's boat went the same distance on all 3 trials and lost to Team A's boat by 7 cm.

   How far did Team B's boat go on each trial? _____

3. Team D's car went the same distance on each of its trials. Team C's car went exactly 1 cm farther in each trial than Team D's car. Team C's car went 543 cm in all.

   How far did Team D's car go on each trial? _____

Practice

4. $5,624 \div 8 =$ _____

5. $8,500 \div 3 =$ _____

6. $4\overline{)9,207}$

7. $5\overline{)3,578}$

Finding Angles of Fraction Circle Pieces

Lesson 8-2

NAME DATE TIME

① Fill in the chart below.

Fraction Circle Piece	Fraction of a Whole	Measure of the Angle
Red	1 whole	
Pink		
Orange		
Yellow		
Dark green		
Light blue		
Dark blue		
Purple		
Light green		

② Name a way you can find the measure of the angles above.

____

____

③ Fill in the chart. Write the sum of the measure of the angles or find a combination of fraction circle pieces with the given sum.

Pink + Orange	
	135°
5 Purples	
1 Orange + 4 Light greens	
	300°
3 Light blues + 2 Light greens	
4 Dark blues + 2 Yellows	
	111°

Measuring Baseball Angles

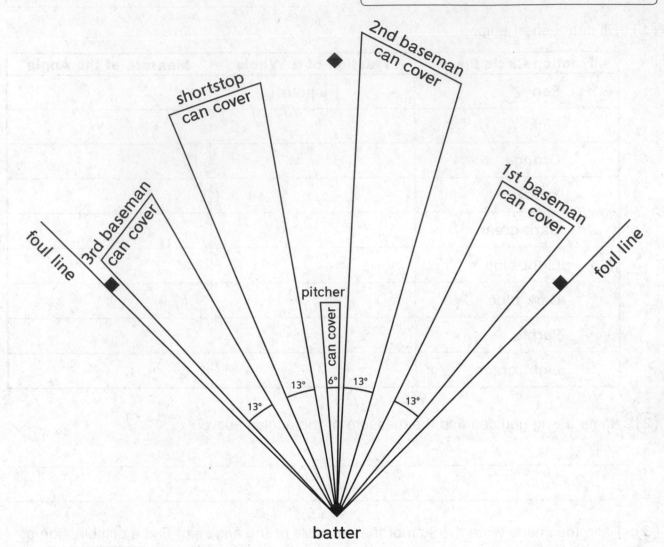

2nd baseman can cover

shortstop can cover

3rd baseman can cover

1st baseman can cover

foul line

foul line

pitcher can cover

3rd baseman

13°   13°   6°   13°   13°

batter

The playing field for baseball lies between the foul lines, which form a 90° angle. Suppose that each of the four infielders can cover an angle of about 13° on a hard-hit ground ball and that the pitcher can cover about 6°. (See the diagram above.)

Source: *Applying Arithmetic*, Usiskin, Z. and Bell, M. © 1983 University of Chicago

How many degrees are left for the batter to hit through? _____

Finding Unknown Angle Measures

Find the missing angle measures. For each problem, write an equation with a letter for the unknown to show how you found your answer.

(1)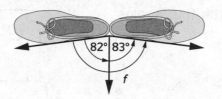

82° 83°

f

f = _____

Equation: _____

(2)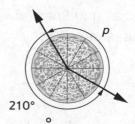

p

210°

p = _____

Equation: _____

(3)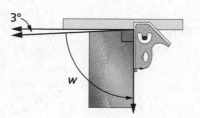

3°

w

w = _____

Equation: _____

(4)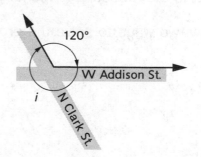

75°   l

l = _____

Equation: _____

(5)

43°   s

s = _____

Equation: _____

(6)

120°

W Addison St.

i

N Clark St.

i = _____

Equation: _____

Practice

(7) $\frac{1}{3} + \frac{2}{3} + \frac{2}{3} =$ _____

(8) $\frac{1}{4} + \frac{3}{4} + \frac{3}{4}$ _____

(9) $\frac{4}{5} + \frac{4}{5} + \frac{3}{5} =$ _____

(10) $\frac{5}{12} + \frac{3}{12} + \frac{7}{12} =$ _____

Pattern-Block Angle Measures

Lynn finds a complete set of pattern blocks. Help Lynn figure out the size of the angles in some of the different shapes.

You can use what you know about squares: Each of the four angles of a square measures 90°.

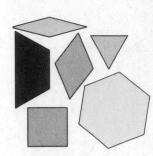

Put some blocks on top of others to fit angles together to figure out the measures of the angles.

① What is the measure of the *small* angle of a white rhombus? _____ °

Draw a picture and explain how you know.

② What is the measure of the *large* angle of a white rhombus? _____ °

Show two ways to solve the problem. Draw pictures and explain how you know.

Pattern-Block Angle Measures (continued)

③ Julie and Perry solved the problem below in different ways.

What is the measure of an angle of a yellow hexagon?
Draw a picture and explain how you know.

Julie's Solution:	**Perry's Solution:**
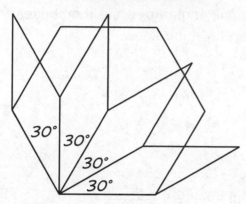	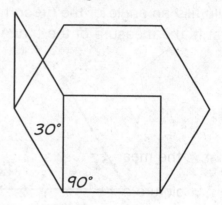

Julie's Solution:

I know that the measure of the hexagon's angle is 120°. The measure of the white rhombus's small angle is 30°. Four angles measuring 30° fit inside the hexagon's angle.

So, 30° + 30° + 30° + 30° = 120°.

Perry's Solution:

I know that the measure of the hexagon's angle is 120° because the measure of the square's angle is 90° and the measure of the white rhombus's small angle is 30°.

So, 90° + 30° = 120°.

Who is correct, Julie, Perry, or both? Write a note to another student explaining your thinking on the back of this page.

Finding Pattern-Block Measures

Molly is using pattern blocks to find angle measures of other pattern blocks. She knows that the measure of the small angle of a white rhombus is 30°.

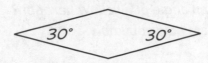

(1) Molly fills an angle of the green triangle with the small angles of white rhombuses. What is the measure of the triangle's angle? Explain how you know.

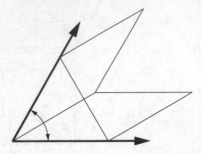

Angle measure: _____

(2) Molly fills a red trapezoid's large angle with angles of the green triangle. What is the measure of the red trapezoid's large angle? Explain how you know.

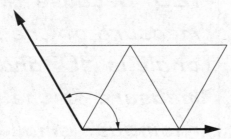

Angle measure: _____

Practice

(3) 5,588 * 3 = _____

(4) 9,037 * 5 = _____

(5) 52 * 94 = _____

(6) 83 * 77 = _____

9-Patch Pattern Pieces

9-Patch Pattern Grid

Creating Symmetric Patterns

Color the traditional 9-patch quilt patterns below so that each has the given number of lines of symmetry. Use a straightedge to draw the lines of symmetry.

① 4 lines of symmetry

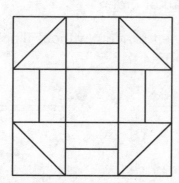

② 2 lines of symmetry

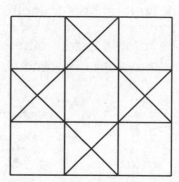

③ 1 line of symmetry

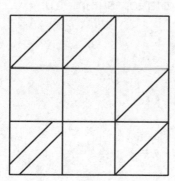

④ 0 lines of symmetry

Rotation Symmetry

A figure has **rotation symmetry** if, when it is rotated less than 360 degrees around a point, the resulting figure exactly matches the original figure. The figure may not be flipped over.

If a figure has rotation symmetry, its **order of rotation symmetry** is the number of different ways it can be rotated to match itself exactly. "No rotation" is counted as one of the ways. So a figure with no rotation symmetry has rotation symmetry of order 1.

A square can be rotated in four different ways and match itself exactly (without flipping it over). Therefore the order of rotation symmetry for a square is 4.

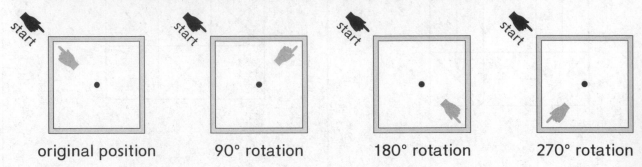

original position   90° rotation   180° rotation   270° rotation

Record the order of rotation symmetry for each figure below.

①

Order of rotation symmetry: _____

②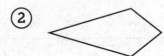

Order of rotation symmetry: _____

③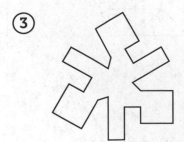

Order of rotation symmetry: _____

④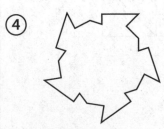

Order of rotation symmetry: _____

Rotation Symmetry

(continued)

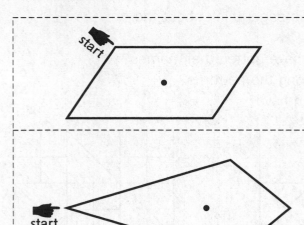

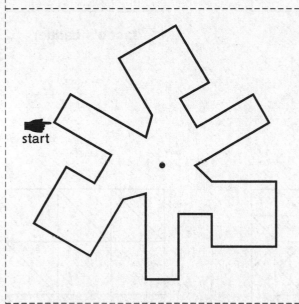

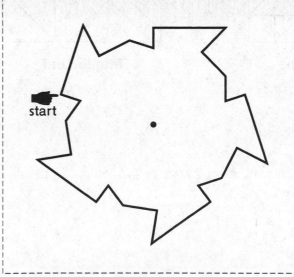

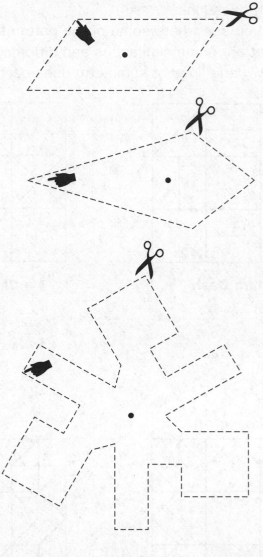

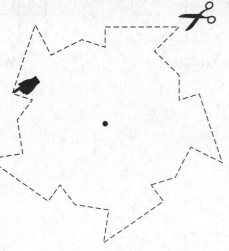

Making a Paper Patchwork Quilt

As you study the traditional 9-Patch Patterns on this page, think about the following questions:

- Do you see where some of the patterns might have gotten their names?
- What are some similarities and differences among the patterns?
- How many lines of symmetry does each pattern have?

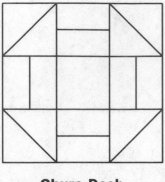

Churn Dash

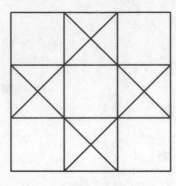

Ohio Star

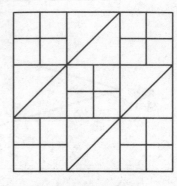

Jacob's Ladder

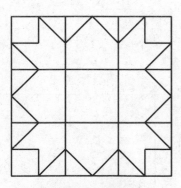

Storm at Sea

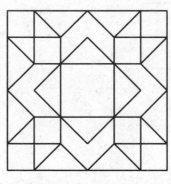

Weather Vane

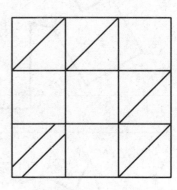

Maple Leaf

Making a Paper
Patchwork Quilt (continued)

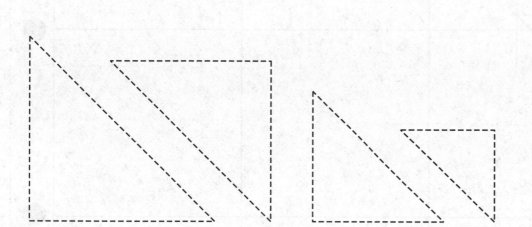

Making a Paper Patchwork Quilt (continued)

Line Symmetry

Use a straightedge to draw the lines of symmetry on each shape.

SRB
238

(1) Draw 2 lines of symmetry.

(2) Draw 6 lines of symmetry.

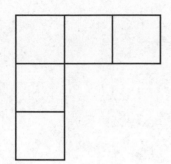

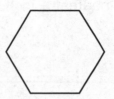

(3) Draw 1 line of symmetry.

(4) Draw 3 lines of symmetry.

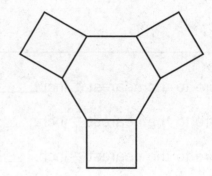

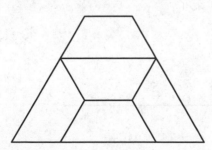

(5) How many lines of symmetry does this shape have? _____

Draw the line(s) of symmetry.

(6) Draw your own shape. Show the lines of symmetry. Be sure your shape includes at least 1 right angle.

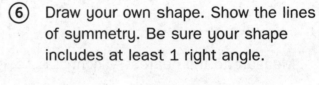

Practice

(7) $6 * \frac{5}{6} =$ _____

(8) $3 * \frac{3}{8} =$ _____

(9) $4 * \frac{7}{10} =$ _____

(10) $6 * \frac{4}{12} =$ _____

315

Measuring to the Nearest $\frac{1}{8}$ Inch

① _____

Measure to the nearest $\frac{1}{8}$ inch. _____ inches

Measure to the nearest $\frac{1}{4}$ inch. _____ inches

② _____

Measure to the nearest $\frac{1}{8}$ inch. _____ inches

Measure to the nearest $\frac{1}{4}$ inch. _____ inches

③ _____

Measure to the nearest $\frac{1}{8}$ inch. _____ inches

Measure to the nearest $\frac{1}{4}$ inch. _____ inches

Measure to the nearest $\frac{1}{2}$ inch. _____ inches

④ _____

Measure to the nearest $\frac{1}{8}$ inch. _____ inches

Measure to the nearest $\frac{1}{4}$ inch. _____ inches

Measure to the nearest $\frac{1}{2}$ inch. _____ inches

⑤ _____

Measure to the nearest $\frac{1}{8}$ inch. _____ inches

Measure to the nearest $\frac{1}{4}$ inch. _____ inches

Measure to the nearest $\frac{1}{2}$ inch. _____ inches

Creating Line Plots
with Fraction Data

Lengths of Objects Data (in inches)

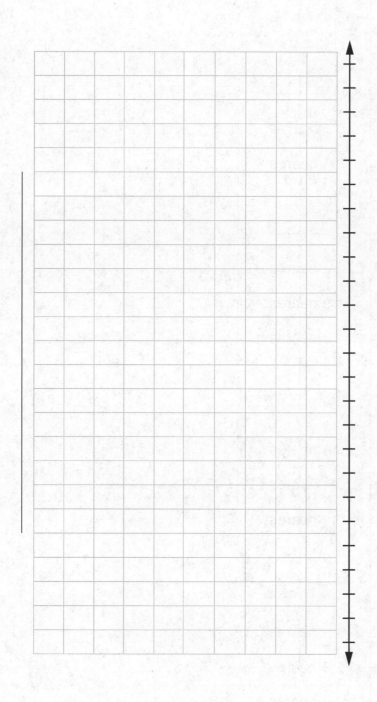

Creating a Greeting Card Line Plot

Lengths of Greeting Cards Data (in inches)

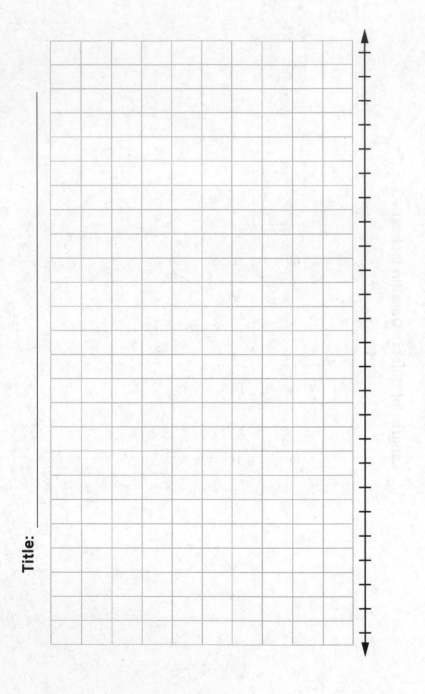

Title:

Designing a Bookcase

Nicholas is building a bookcase. To help with the design, he measured the height of each of his books to the nearest $\frac{1}{8}$ inch. His measurements are given below.

$6\frac{1}{2}$, $9\frac{1}{4}$, $7\frac{1}{8}$, $7\frac{1}{2}$, 8, $6\frac{7}{8}$, $9\frac{1}{4}$, $9\frac{1}{4}$, $9\frac{1}{4}$, $9\frac{1}{4}$, $9\frac{1}{4}$, $8\frac{1}{4}$, 8, $8\frac{1}{4}$, $8\frac{3}{8}$,

$6\frac{1}{2}$, $7\frac{1}{8}$, 9, $6\frac{7}{8}$, $9\frac{3}{8}$, $6\frac{7}{8}$, $7\frac{1}{2}$, 8, $8\frac{1}{4}$, $9\frac{1}{4}$, $6\frac{7}{8}$, $6\frac{7}{8}$, $8\frac{1}{4}$, $8\frac{1}{4}$, $8\frac{1}{4}$

Plot the data set on the line plot below.

Book Heights

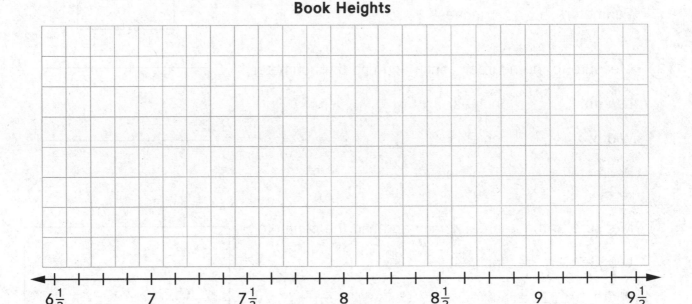

Height (inches)

Use the completed line plot to answer the questions below.

1. What is the difference in height between the tallest and shortest books? _____ in.

2. Nicholas wants the space between the shelves to be $\frac{7}{8}$ inch taller than his tallest book.

   a. How far apart should he make the shelves? _____ in.

   b. If the thickness of the wood he uses for the shelves is $\frac{5}{8}$ inch, what will be the total height of each shelf? (*Hint:* The total height is the thickness of one piece of wood plus the distance between shelves.) _____ in.

Practice

3. $8,207 \div 7 \rightarrow$ _____

4. $7,109 \div 8 \rightarrow$ _____

Measuring to Find Perimeter

Measure the length and the width, and then use a formula to find the perimeter. Show your work on the back of this paper.

SRB
162, 175, 200

① Measure to the nearest $\frac{1}{4}$ inch and find the perimeter.

Length: _____ inch(es)

Width: _____ inch(es)

Perimeter: _____ inches

② Measure to the nearest $\frac{1}{4}$ inch and find the perimeter.

Length: _____ inch(es)

Width: _____ inch(es)

Perimeter: _____ inches

③ Measure to the nearest $\frac{1}{8}$ inch and find the perimeter.

Length: _____ inch(es)

Width: _____ inch(es)

Perimeter: _____ inches

④ Measure to the nearest $\frac{1}{8}$ inch and find the perimeter.

Length: _____ inch(es)

Width: _____ inch(es)

Perimeter: _____ inches

⑤ Measure to the nearest $\frac{1}{8}$ inch and find the perimeter.

Length: _____ inch(es)

Width: _____ inch(es)

Perimeter: _____ inches

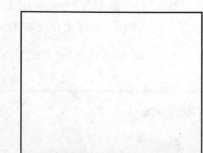

Perimeters and Missing Measures

Use a formula to find the perimeter of each rectangle. Show your work in the space provided.

---

① Length = $3\frac{3}{6}$ yd

Width = $\frac{1}{6}$ yd

Perimeter: _____ yd

② Length = $5\frac{1}{12}$ ft

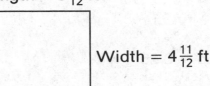

Width = $4\frac{11}{12}$ ft

Perimeter: _____ ft

For each rectangle, find the unknown side measure.

---

③ Perimeter: $\frac{74}{100}$ kilometer

Length = $\frac{25}{100}$ km

Width = _____ km

④ Perimeter: 10 inches

Length = $4\frac{3}{8}$ in.

Width = _____ in.

---

⑤ Perimeter: $12\frac{8}{10}$ centimeters

Length = $4\frac{1}{10}$ cm

Width = _____ cm

Try This

⑥ Perimeter: $16\frac{1}{2}$ ft

Length = _____ ft

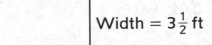

Width = $3\frac{1}{2}$ ft

Practice

⑦ $2 * \frac{2}{3} =$ _____

⑧ $5 * \frac{3}{4} =$ _____

⑨ $9 * \frac{4}{5} =$ _____

⑩ $8 * \frac{6}{12} =$ _____

321

Adding Tenths and Hundredths

Model each problem with base-10 blocks. Change the decimals to fractions and solve. Use a long │ to represent 0.1, or $\frac{1}{10}$. Use a cube ▪ to represent 0.01, or $\frac{1}{100}$. Write an equation to show your work.

SRB
166-167

① 0.55 + 0.25

Answer: _____

Equation: _____

② 0.34 + 0.17

Answer: _____

Equation: _____

③ 0.5 + 0.16

Answer: _____

Equation: _____

④ 0.33 + 0.4

Answer: _____

Equation: _____

⑤ 0.02 + 0.8

Answer: _____

Equation: _____

Designing a Baseball Cap Rack

Olivia wants to design and build two horizontal racks with pegs to display her baseball cap collection. She has 12 caps, but her sister suggested that she leave space for as many caps as possible so that she can add to her collection. After measuring her caps, Olivia decided to place the pegs 0.2 m apart. Each peg is 1 cm wide. In order to fit on her wall, each rack can be no longer than 1.6 meters long.

Help Olivia design the racks.

(1) What is the greatest number of pegs each rack can have? _____

(2) The total length of one rack will be _____ m.

(3) The first peg will be _____ m from the edge of the rack.

(4) Sketch one of the two racks in the space below. Label your measurements.

(5) Write number sentences that show how many pegs could fit on one rack.

_____

_____

(6) Could there be 9 pegs on the rack? Explain your answer.

_____

_____

Designing a Baseball Cap Rack (continued)

At the lumberyard, Olivia discovered that she could save money by purchasing leftover pieces of wood. None of the pieces were long enough, but she decided to glue pieces together to make the lengths she needed. She had the following lengths of board from which to choose:

0.7 m	0.85 m
0.35 m	0.3 m
0.20 m	0.9 m
0.8 m	0.55 m
0.75 m	0.15 m

(7) Can Olivia use these pieces to create two racks of the length she planned? Explain why or why not. Show your work.

_____

_____

_____

_____

(8) Pegs come in two different types of packages: a 5-pack for $3.70 or a 2-pack for $1.99. Explain how Olivia can purchase the pegs for her racks, spending as little money as possible.

_____

_____

_____

_____

Solving Olympic Number Stories

The difference between a gold medal and no medal in the Olympics can come down to hundredths of a second or a point. Solve the following number stories about the 2014 Winter Olympics.

(1) The top three medal winners in men's ski jumping had the following total points: 1,041.1 points, 1,038.4 points, and 1,024.9 points. How many points total did these 3 skiers earn all together?

Number model with unknown: _____

Answer: _____ points

(2) The first-place women's figure skater had a total score of 224.59 points. The tenth-place skater had 174.53 points. How many more points did the first-place skater score than the tenth-place skater?

Number model with unknown: _____

Answer: _____ points

(3) The men's figure skating medal winners' total scores were: Gold: 280.09; Silver: 275.62; Bronze: 255.10. How much higher was the gold medal winner's score than the bronze medal winner's?

Number model with unknown: _____

Answer: _____ points

(4) In the 1948 Winter Olympics in St. Moritz, Switzerland, the medals were 3.8 mm thick. In the 2014 Winter Olympics in Sochi, Russia, they were 10 mm thick.

a. How tall was a stack of 8 of the 1948 medals?

Number model with unknown: _____

Answer: _____ mm

b. What is the total difference in height between a stack of 8 of the medals from 1948 and a stack of 8 of the medals from 2014?

Number model with unknown: _____

Answer: _____ mm

Decimal Number Stories

Solve each number story. Write your answer as a decimal.
Show how you found your answer.

SRB
166-167

(1) An Olympic men's shot put weighs 7.26 kilograms. An Olympic women's shot put weighs 4 kilograms. How much more does the men's shot put weigh than the women's shot put?

_____ kilograms

(2) The recipe for homemade glue calls for 0.5 liter of skim milk, 0.09 liter of vinegar, and 0.06 liter of water. When you combine the ingredients, how much liquid will you have?

_____ liter

(3) Ben cut a piece of string 11.4 cm long. Then he cut 3.6 cm off of it. How long is the string now?

_____ cm

Try This

(4) What is the answer to Problem 3 in milliliters? _____ milliliters

Practice

(5) $3,579 * 4 =$ _____

(6) $2,904 * 6 =$ _____

(7) $36 * 56 =$ _____

(8) $47 * 72 =$ _____

Finding the Area

Find the area of the rectangles.

SRB 175, 204

①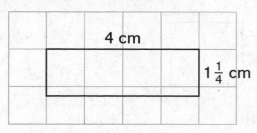

Number model with unknown:

_____

Answer: _____ square cm

②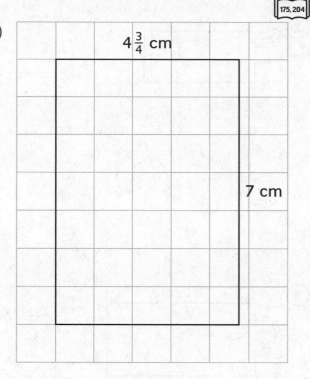

Number model with unknown:

_____

Answer: _____ square cm

③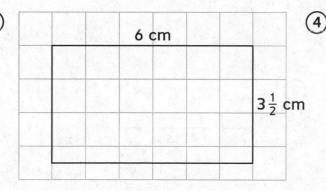

Number model: _____

Answer: _____ square cm

④

Number model: _____

Answer: _____ square cm

Finding Area and Unknown Side Lengths

Find the area or unknown side length for each rectangle.

(1)
9 cm
4.3 cm

Area: _____ square cm

(2)
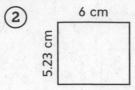
6 cm
5.23 cm

Area: _____ square cm

(3)
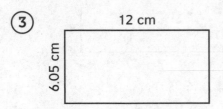
12 cm
6.05 cm

Area: _____ square cm

(4)
8 cm

Area = 40.64 sq cm

Width: _____ cm

(5)
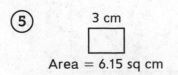
3 cm

Area = 6.15 sq cm

Width: _____ cm

(6)
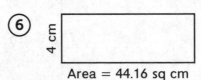
4 cm

Area = 44.16 sq cm

Width: _____ cm

Measuring to Find Area

① Measure to the nearest $\frac{1}{2}$ inch and find the area.

Length: _____ inches   Width: _____ inches   Area: _____ square inches

② Measure to the nearest $\frac{1}{4}$ inch and find the area.

Length: _____ inches   Width: _____ inches   Area: _____ square inches

③ Measure to the nearest $\frac{1}{8}$ inch and find the area.

Length: _____ inches   Width: _____ inches   Area: _____ square inches

329

Area and Perimeter

Solve the problems below.

SRB
175-204

1. The Murphy family bought two rectangular dog beds for their pets. Fluffy's bed was 3 feet by $1\frac{9}{12}$ feet. Pete's bed was 4 feet by $2\frac{4}{12}$ feet.

   a. How much more area does Pete's bed have than Fluffy's?

   Answer: _____ square feet

   b. What is the perimeter of Pete's bed? Answer: _____ feet

2. The Cho family bought two rectangular cat beds for their cats. George's bed is 2 feet by $1\frac{2}{12}$ feet. Sammie's bed is 2 feet by $1\frac{7}{12}$ feet.

   a. What is the total area of these two beds? Answer: _____ square feet

   b. What is the perimeter of George's bed? Answer: _____ feet

3. Perimeter: $12\frac{2}{10}$ inches

   x inches

   $2\frac{1}{10}$ inches

   Area: _____ square inches

4. Area: $9\frac{3}{8}$ square feet

   x feet

   3 feet

   Width: _____ feet

Practice

5. $\frac{5}{6} - \frac{1}{6} =$ _____

6. $\frac{8}{8} - \frac{3}{8} =$ _____

7. $\frac{9}{10} - \frac{5}{10} =$ _____

8. $\frac{11}{12} - \frac{5}{12} =$ _____

Movie Lengths

NAME   DATE   TIME

Approximate Lengths of Animated Movies	
Movie Title	Length (in hours)
A Christmas Carol	$1\frac{2}{3}$
Cars	$1\frac{11}{12}$
Cars 2	$1\frac{11}{12}$
Finding Nemo	$1\frac{2}{3}$
Frozen	$1\frac{2}{3}$
Ice Age	$1\frac{5}{12}$
Kung Fu Panda	$1\frac{1}{2}$
Madagascar	$1\frac{5}{12}$
Monsters, Inc.	$1\frac{1}{2}$
Ratatouille	$1\frac{5}{6}$
Shrek	$1\frac{1}{2}$
Tangled	$1\frac{2}{3}$
The Incredibles	$1\frac{11}{12}$
The Lion King	$1\frac{1}{2}$
Toy Story 3	$1\frac{3}{4}$
Up	$1\frac{5}{12}$
Wreck-It Ralph	$1\frac{5}{6}$

Practicing for an Audition

The Gateway Acting Club is holding auditions for a new play. The director suggested that club members practice a song at home to prepare for the auditions.

SRB
160, 162, 175

Solve the number stories below. Use equations or drawings to show how you solved each problem.

(1) Geoff practiced for $\frac{1}{4}$ hour every day for 2 weeks. How much time did he practice all together?

Answer: _____ hour(s)

(2) Jocelyn practiced for $\frac{1}{5}$ hour each day for a week and $\frac{2}{5}$ hour each day the next week. How much time did Jocelyn spend practicing in all?

Answer: _____ hour(s)

(3) Barrie practiced for $\frac{1}{10}$ hour every day the first week and $\frac{3}{10}$ hour every day the second week. How much time did Barrie practice all together?

Answer: _____ hour(s)

(4) Sydney practiced for $\frac{1}{6}$ hour on 5 days, $\frac{3}{6}$ hour on 2 days, and $\frac{2}{6}$ hour on 7 days. How much time did Sydney spend practicing in all?

Answer: _____ hour(s)

(5) Put the actors in order of the amount of time they practiced from most to least.

_____ , _____ , _____ , _____

Using Doghouse Dimensions

Dan and Diane's Doghouse Dynasty builds doghouses to order. They can change the length and width for doghouses, but they always build them to have the same height. Solve the number stories about doghouses built to certain widths and lengths based on the information given in the table. Use drawings or equations to show how you solved each problem.

SRB
162, 175

Custom Doghouse Dimensions		
Size	Length (in feet)	Width (in feet)
Extra small	$3\frac{1}{4}$	$1\frac{1}{3}$
Small	$3\frac{1}{2}$	$1\frac{1}{2}$
Medium	4	$1\frac{3}{4}$
Large	$4\frac{1}{4}$	$1\frac{5}{6}$
Extra large	$4\frac{5}{6}$	2

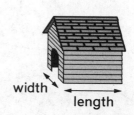

width   length

(1) Mrs. Swift ordered 3 medium-size doghouses. What will their combined width be?

_____ feet

(2) Kisa's Kennel has a space that is 18 feet wide in which they want to place doghouses side by side. If they order 5 small and 4 medium doghouses, will they all fit in the space? _____

Practice

(3) $2 * \frac{3}{6} =$ _____

(4) $5 * \frac{7}{10} =$ _____

(5) $9 * \frac{6}{100} =$ _____

(6) $7 * \frac{8}{12} =$ _____

Showing Liquid Measurements

Use the "Big G" to help you find the answers.

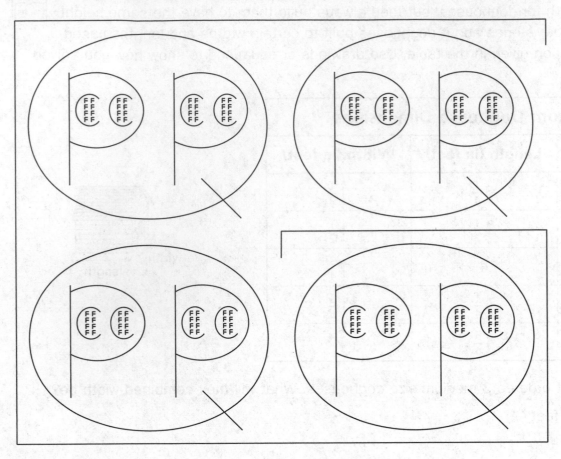

① 1 gal = _____ qt _____ pt _____ c	**②** $\frac{1}{2}$ gal = _____ qt _____ pt _____ c	**③** $\frac{1}{4}$ gal = _____ qt _____ pt _____ c
④ 2 gal = _____ qt _____ pt _____ c	**⑤** 3 gal = _____ qt _____ pt _____ c	**⑥** $1\frac{1}{2}$ gal = _____ qt _____ pt _____ c

Converting Units of Liquid Measure

Cut out the measurements in Columns B and C and put them in separate piles. For each measurement in Column A, find one equivalent measurement from Column B and one from Column C. Paste the equivalent measurements in a row.

Column A	Column B	Column C
$3\frac{1}{2}$ gallons	4 cups	736 fl oz
$3\frac{1}{2}$ qt	18 pt	36 cups
$6\frac{1}{4}$ gal	3 cups	16 fl oz
$2\frac{1}{4}$ gallons	1 pint	32 fl oz
$\frac{1}{4}$ gallon	27 c	112 fl oz
$\frac{1}{2}$ quart	50 pints	28 pints
$1\frac{1}{2}$ pt	46 pt	80 fl oz
$2\frac{1}{2}$ quarts	14 cups	216 fl oz
$5\frac{3}{4}$ gallons	5 pints	50 c + 400 fl oz
$6\frac{3}{4}$ quarts	14 qt	24 fl oz

335

Solving Liquid Measurement Puzzles and Problems

Fill in the blanks below.

① **Rule:** _____

in	out
$\frac{1}{4}$ cup	$\frac{2}{4}$ cup
	$\frac{3}{4}$ cup
$\frac{3}{4}$ cup	1 cup
	$2\frac{3}{4}$ cups

② **Rule:** _____

in	out
$\frac{3}{4}$ gallon	$\frac{1}{4}$ gallon
	1 gallon
$3\frac{1}{2}$ gallons	3 gallons
	$4\frac{1}{2}$ gallons

③ **Rule:** _____

in	out
$\frac{3}{8}$ cup	$\frac{6}{8}$ cup
2 cups	$2\frac{3}{8}$ cups
	3 cups
	$5\frac{6}{8}$ cups

④ **Rule:** _____

in	out
$\frac{1}{2}$ pint	$1\frac{1}{2}$ pints
1 pint	
	6 pints
$6\frac{7}{8}$ pints	

As part of a health project, 4 families kept track of how much liquid besides water they drank in one week. Use the information in the table to solve the following number story.

Gallons of Liquid Consumed in One Week				
Family Name	Milk	Apple Juice	Orange Juice	Tomato Juice
Aniciete	$6\frac{3}{4}$	$2\frac{3}{4}$	$2\frac{3}{4}$	$\frac{1}{4}$
Baker	$1\frac{1}{2}$	$2\frac{3}{8}$	$2\frac{3}{8}$	$\frac{3}{8}$
Shakur	$2\frac{1}{8}$	$3\frac{3}{8}$	$1\frac{1}{2}$	$\frac{1}{8}$
Vega	$5\frac{1}{4}$	$\frac{3}{4}$	$2\frac{1}{4}$	$\frac{1}{8}$

⑤ How much liquid besides water did the Vega family drink in 3 weeks?

Number model with unknown: _____

Answer: _____ gallon(s), or _____ quart(s), or _____ pint(s), or _____ cup(s), or _____ fluid ounces

Liquid Measurement and Fractions

Complete the "What's My Rule?" tables and state the rules.

(1) Rule: _____

in (gallons)	out (pints)
2	16
$3\frac{1}{2}$	
	48
$7\frac{1}{4}$	
	80

(2) Rule: _____

in (quarts)	out (cups)
3	12
$4\frac{1}{2}$	
	32
$9\frac{3}{4}$	
$12\frac{1}{4}$	

Use this recipe for a Creamsicle Smoothie to solve the problems below.

$\frac{3}{4}$ cup orange juice   4 fluid ounces cold water   1 cup vanilla ice cream

Combine all ingredients.

(3) **a.** Will this recipe fit in a glass that holds 24 fluid ounces? _____

   Explain your thinking. _____

   _____

   b. About how many more cup(s) of smoothie could fit in the glass? _____ cup(s)

   c. Frank wants to triple the recipe. How much of each ingredient will he need?

   _____ orange juice

   _____ cold water

   _____ vanilla ice cream

   d. After tripling the recipe, how much smoothie will Frank have? _____ fluid ounces

Practice

(4) 3,560 ÷ 3 → _____

(5) 9,295 ÷ 5 → _____

(6) 7)8,210

(7) 9)4,671

337

Understanding Fractions of Pounds

OZ	OZ	OZ	OZ
OZ	OZ	OZ	OZ
OZ	OZ	OZ	OZ
OZ	OZ	OZ	OZ

Understanding Ounces

Circle the correct answer for each problem below. Use your answers to solve the puzzle at the bottom of the page.

SRB
191-192,
196-197

(1) How many fluid ounces in 4 pints of lemonade?

M. 56 fl oz

N. 60 fl oz

O. 64 fl oz

P. 68 fl oz

(2) How many ounces in $2\frac{1}{2}$ pounds of potatoes?

P. 32 ounces

Q. 38 ounces

R. 40 ounces

S. 44 ounces

(3) How many ounces in $5\frac{3}{4}$ pounds gold?

J. 80 ounces

K. 88 ounces

L. 92 ounces

M. 96 ounces

(4) How many ounces in $3\frac{3}{4}$ pounds of rocks?

D. 40 ounces

E. 48 ounces

F. 56 ounces

G. 60 ounces

(5) How many fluid ounces in $\frac{7}{8}$ cup of water?

A. 4 fl oz

B. 5 fl oz

C. 6 fl oz

D. 7 fl oz

(6) How many fluid ounces in $4\frac{1}{2}$ quarts of oil?

W. 112 fl oz

X. 128 fl oz

Y. 144 fl oz

Z. 160 fl oz

(7) How many fluid ounces in $1\frac{1}{4}$ gallons of milk?

D. 148 fl oz

E. 150 fl oz

F. 160 fl oz

G. 174 fl oz

(8) How many fluid ounces in $2\frac{1}{4}$ cups of tea?

A. 18 fl oz

B. 24 fl oz

C. 32 fl oz

D. 40 fl oz

(9) How many ounces in $2\frac{7}{8}$ pounds of sand?

M. 40 ounces

N. 46 ounces

O. 48 ounces

P. 54 ounces

Write the letter of each answer above its problem number to complete the sentence below.

___ ___ ___ ___ ___ ___ ___ ___ ___
 5   2   8   4   1   9   7   3   6

According to the Smithsonian Institution, the fastest flying insect is the

_____, which can fly 35 mph.

Planning a Cookout

The Whispering Lakes Neighborhood Association is having a hamburger cookout. Each family can choose whether to order the hamburgers or bring their own. Use the information in the table to solve the number stories. Use drawings, tables, or equations to show what you did.

Size of Hamburger	Weight of One Hamburger Patty (lb)
Small	$\frac{1}{8}$
Medium	$\frac{1}{4}$
Large	$\frac{1}{2}$
Jumbo	$\frac{3}{4}$
King of the Burgers	$1\frac{1}{2}$

SRB
190-191

① a. What is the combined weight of 1 of each size hamburger?

_____ pounds

b. How many ounces is that?

_____ ounces

c. Mrs. Ward found 80-ounce packages of hamburger on sale. If she needs to make 2 of each size hamburger, how many packages of meat will she need to buy?

_____ packages

② The Finch family ordered 2 small hamburgers, 1 medium hamburger, and 1 jumbo hamburger. How many pounds of hamburger meat does the neighborhood association need to buy for this family?

_____ pounds

Practice

③ $5,107 * 3 =$ _____

④ $4,794 * 6 =$ _____

⑤ $74 * 29 =$ _____

⑥ $93 * 48 =$ _____

Number-Tile Computations

Cut out the 0–9 number tiles at the bottom of the page. Use them to help you solve the problems. Each of the 20 tiles can only be used once.

SRB 92-95

① Use odd-numbered tiles 1, 3, 5, 7, and 9 to make the largest sum.

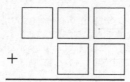

② Use even-numbered tiles 0, 2, 4, 6, and 8 to make the smallest difference.

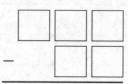

③ Use number tiles 0, 4, 6, and 8 to make the largest product.

④ Use number tiles 1, 2, 5, and 7 to make the smallest whole-number quotient. The answer may have a remainder.

☐☐☐ / ☐ → _____

⑤ Answer the following questions using only the unused tiles and any operation. Write number sentences to show your work.

a. What is the largest answer you can find? _____

☐ ___ ☐ = ___

b. What is the smallest answer you can find? _____

☐ ___ ☐ = ___

Practice

⑥ $4\frac{3}{5} + 3\frac{4}{5} =$ _____

⑦ $1\frac{5}{8} + 3\frac{5}{8} =$ _____

⑧ $2\frac{9}{12} + 4\frac{5}{12} =$ _____

⑨ $5\frac{89}{100} + 5\frac{92}{100} =$ _____

0	1	2	3	4	5	6	7	8	9
0	1	2	3	4	5	6	7	8	9

341

Record Sheet for Broken Calculator

Player 1 Name:

Target Number	Number Model	Score
	Total Score	

Player 2 Name:

Target Number	Number Model	Score
	Total Score	

Player 3 Name:

Target Number	Number Model	Score
	Total Score	

Solving a Broken-Calculator Dilemma

Caleb wanted to find the answer to $z + 643 = 1,210$. His teacher said, "Use your calculator, but pretend the subtraction key is broken." Caleb used a guess-and-check strategy to solve the problem and organized his work in a table like this one.

Broken key: $\boxed{-}$
To solve: $z + 643 = 1,210$

$600 + 643 = 1,243$	Too much
$550 + 643 = 1,193$	Too little
$560 + 643 = 1,203$	Closer
$567 + 643 = 1,210$	Got it!

Solve each problem on your calculator without using the "broken" key. Only one key is broken in each problem. Record your steps.

① Broken key: $\boxed{+}$
To solve: $d - 574 = 1,437$

② Broken key: $\boxed{*}$
To solve: $w / 15 = 8$

Make up one for your partner to solve.

③ Broken key: $\boxed{÷}$
To solve: $s * 48 = 2,928$

④ Broken key: $\boxed{}$
To solve:

Many Names for Numbers

Write five names in each box below. Use as many different kinds of numbers (such as whole numbers, fractions, decimals) and different operations (+, −, *, ÷) as you can. **SRB** 33

①
9,990

②
32.68

Make up your own name-collection boxes.

③

④

Practice

⑤ $5\frac{1}{4} - 1\frac{3}{4} =$ _____

⑥ $4\frac{3}{10} - 2\frac{7}{10} =$ _____

⑦ $6\frac{7}{12} - 3\frac{11}{12} =$ _____

⑧ $8\frac{1}{6} - 4\frac{5}{6} =$ _____

End-of-Year Family Letter

Congratulations!

By completing *Fourth Grade Everyday Mathematics*, your child has accomplished a great deal. Thank you for all of your support this year.

This Family Letter is a resource to use throughout your child's vacation. It includes an extended list of "Do-Anytime Activities," directions for games that can be played at home, a list of mathematics-related books to check out over vacation, and a sneak preview of what your child will be learning in *Fifth Grade Everyday Mathematics*. Enjoy your vacation!

Do-Anytime Activities

Mathematics means more to everyone when it is rooted in real-life situations. To help your child review many of the concepts he or she has learned in fourth grade, we suggest the following activities for you to do together over the break. These activities will not only help to prevent your child from forgetting content, but they will also help prepare him or her for *Fifth Grade Everyday Mathematics*.

1. Practice multiplication and division facts to maintain fluency.

2. Convert measurements in real-world contexts. For example, at the grocery store ask, "How many quarts are in this gallon of milk?"

3. Have your child practice multidigit multiplication and division using the algorithms with which he or she is most comfortable.

4. Look at advertisements and compare sale prices to original prices. Use a calculator to find unit prices to determine possible savings.

Building Skills through Games

The following section lists rules for games that can be played at home. You will need a deck of number cards, which can be made from index cards or by modifying a regular deck of cards as follows:

A regular deck of playing cards includes 54 cards (52 regular cards plus 2 jokers). Use a permanent marker to write on the cards or a ballpoint pen to write on pieces of white adhesive labels to mark some of the cards:

- Mark each of the four aces with the number "1."
- Mark each of the four queens with the number "0."
- Mark each of the four jacks and the four kings with one of the numbers from 11–18.
- Mark the two jokers with the numbers 19 and 20.

Name That Number

Materials   1 set of cards. See above for directions to make this set.

Players   2 or 3

Object of the Game   To collect the most cards

Directions

1. Shuffle the cards and deal five cards to each player. Place the remaining cards number-side down. Turn over the top card and place it beside the deck. This is the **target number** for the round.

2. Players try to match the target number by adding, subtracting, multiplying, or dividing the numbers on as many of their cards as possible. A card may be used only once.

3. Players write their solutions on a sheet of paper or a slate. When players have written their best solutions, they:

    • Set aside the cards they used to name the target number.

    • Replace used cards by drawing new cards from the top of the deck.

    • Put the old target number on the bottom of the deck.

    • Turn over a new target number and play another hand.

4. Play continues until there are not enough cards left to replace all of the players' cards. The player who sets aside more cards wins the game.

Example: Target number: 16   A player's cards: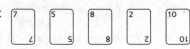

Some possible solutions:

$10 + 8 - 2 = 16$ (*three cards used*)

$7 * 2 + 10 - 8 = 16$ (*four cards used*)

$8 / 2 + 10 + 7 - 5 = 16$ (*all five cards used*)

The player sets aside the cards used to make a solution and draws the same number of cards from the top of the deck.

Top-It Games

Materials   Number cards 1–9 (4 of each) as described above
        1 calculator (optional)

Players   2 to 4

Skills   Addition, Subtraction, and Multiplication

Object of the Game   To collect the most cards

Addition Top-It
Directions

1. Shuffle the cards and place them number-side down on the table.

2. Each player takes eight cards, forms two 4-digit numbers, and finds the sum. Players should carefully consider how they form their numbers, because different arrangements lead to different sums. For example, 7,431 + 5,269 has a greater sum than 1,347 + 2,695. The player with the largest sum takes all the cards. In case of a tie, each player turns over eight more cards and calls out the sum. The player with the largest sum takes all the cards from both rounds.

3. Check answers, using a calculator if necessary.

4. The game ends when there are not enough cards left for each player to have another turn.

5. The player with the most cards wins.

Subtraction Top-It
Directions

1. Shuffle the cards and place the deck number-side down on the table.

2. Each player takes eight cards, forms two 4-digit numbers, and finds the difference. Players should carefully consider how they form their numbers, because different arrangements lead to greater differences. For example, 7,431 − 5,269 has a smaller difference than 7,431 − 2,695. The player with the largest difference takes all the cards. In case of a tie, each player turns over eight more cards and calls out the difference. The player with the largest difference takes all the cards from both rounds.

3. Check answers, using a calculator if necessary.

4. The game ends when there are not enough cards left for each player to have another turn.

5. The player with the most cards wins.

Multiplication Top-It
Directions

1. Shuffle the cards and place them number-side down on the table.

2. Each player turns over four cards, forms two 2-digit numbers, and finds the product. Players should carefully consider how they form their numbers, because different arrangements lead to different products. For example, 74 * 52 has a greater product than 47 * 25. The player with the largest product takes all the cards. In case of a tie, each player turns over four more cards and calls out the product. The player with the largest product takes all the cards from both rounds.

3. Check answers, using a calculator if necessary.

4. The game ends when there are not enough cards left for each player to have another turn.

5. The player with the most cards wins.

Vacation Reading with a Mathematical Twist

Books can contribute to students' learning by representing mathematics in a combination of real-world and imaginary contexts. The titles listed below were recommended by teachers who use *Everyday Mathematics* in their classrooms. They are organized by mathematical topic. Visit your local library and check out these and other mathematics-related books with your child.

Operations and Algebraic Thinking

A Remainder of One by Elinor J. Pinczes

17 Kings and 42 Elephants by Margaret Mahy

Anno's Magic Seeds by Mitsumasa Anno

Pattern by Henry Pluckrose

The Grapes of Math by Greg Tang

Numeration and Operations in Base-Ten

If the World Were a Village by David J. Smith

The Doorbell Rang by Pat Hutchins

The Man Who Counted: A Collection of Mathematical Adventures by Malba Tahan

The Grizzly Gazette by Stuart J. Murphy

Numeration and Operations: Fractions

Fraction Fun by David A. Adler

Working with Fractions by David Adler

Full House by Dayle Ann Dodds

Funny & Fabulous Fraction Stories by Dan Greenberg

Civil War Recipes: Adding and Subtracting Simple Fractions by Lynn George

Music Math: Exploring Different Interpretations of Fractions by Kathleen Collins

My Half Day by Doris Fisher and Dani Sneed

The Wishing Club by Donna Jo Napoli

Measurement and Data

How Tall, How Short, How Faraway by David A. Adler

Is a Blue Whale the Biggest Thing There Is? by Robert E. Wells

Math Curse by Jon Scieszka and Lane Smith

Counting on Frank by Rod Clement

Spaghetti and Meatballs for All! by Marilyn Burns

Geometry

The Greedy Triangle by Marilyn Burns

Grandfather Tang's Story by Ann Tompert

Sweet Clara and the Freedom Quilt by Deborah Hopkinson

Whale of a Tale by Barbara Pearl

Zachary Zormer, Shape Transformer by Joanne Reisberg

Looking Ahead: *Fifth Grade Everyday Mathematics*

Next year, your child will . . .

- Continue to explore and practice whole-number operations, including the use of exponents, and work with larger numbers.

- Expand skills with decimals and fractions, includeing using all four operations.

- Investigate methods for solving problems using mathematics in everyday situations.
  - Graph points on coordinate planes to solve real-world mathematical problems
  - Work with number lines, times, dates, and rates
  - Collect, organize, describe, and interpret numerical data

- Analyze patterns and relationships.

- Further explore the properties, relationships, and measurement of 2-dimensional objects and begin to work with 3-dimensional objects.

- Understand the concepts of volume.

Again, thank you for all of your support this year. Have fun increasing your own understanding of mathematics while continuing your child's mathematical learning!

Teaching Aid Masters

Compact Place-Value Flip Book

NAME   DATE   TIME

1. Cut each page along the dashed lines. Do NOT cut any of the solid lines.

2. Cut along the vertical dashed lines to separate the digits.

3. Assemble the pages in order.

4. Staple the assembled book. (Ask your teacher for help if you need it.)

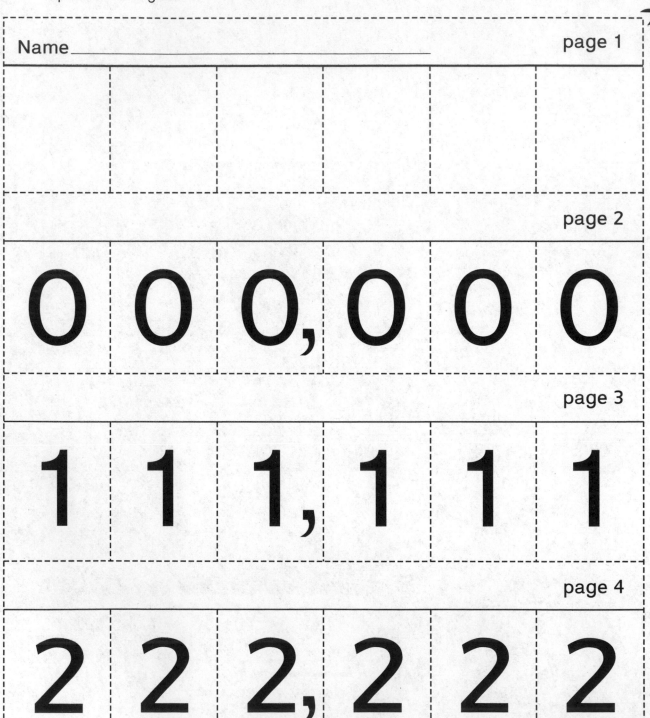

Name _____   page 1

                                                             page 2

0 0 0,0 0 0

                                                             page 3

1 1 1,1 1 1

                                                             page 4

2 2 2,2 2 2

Compact Place-Value
Flip Book (cont.)

NAME   DATE   TIME

page 5

3 3 3, 3 3 3

page 6

4 4 4, 4 4 4

page 7

5 5 5, 5 5 5

TA3

Compact Place-Value
Flip Book (cont.)

page 8

6 6 6,6 6 6

page 9

7 7 7,7 7 7

page 10

8 8 8,8 8 8

TA4

Compact Place-Value
Flip Book (cont.)

page 11

9 9 9,9 9 9

page 12

Hundred-Thousands	Ten-Thousands	Thousands	Hundreds	Tens	Ones

Number-Grid Puzzles

NAME   DATE   TIME

1. Find the missing numbers.

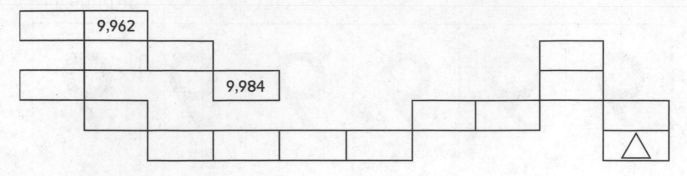

    a. △ = _____   **b.** Explain how you found △.

_____

_____

_____

2. Below is a number-grid puzzle cut from a different number grid.
   Figure out the pattern and use it to fill in the missing numbers.

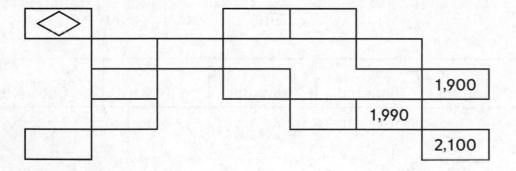

    a. ◇ = _____   **b.** Explain how you found ◇.

_____

_____

_____

    c. Describe how this number grid is different from number grids you have used before.

_____

_____

Array Grid Paper

Reengagement Planning Form

Common Core State Standard (CCSS):	
Goal for Mathematical Practice (GMP):	
Strengths and understandings:	
Weaknesses and misconceptions:	

Planning the Reengagement Discussion

Issue to address	Work samples that illustrate this issue	Questions to ask about the sample student work

Measurement Conversion

_____ to _____

_____ to _____

_____ to _____

_____ to _____

_____ to _____

_____ to _____

_____ to _____

_____ to _____

Dot Paper

Grid Paper (cm)

TA11

Isometric Dot Paper

NAME　　　　　　　　DATE　　TIME

TA12

Multiplication/Division Facts Table

NAME   DATE   TIME

*, /	1	2	3	4	5	6	7	8	9	10
1	1	2	3	4	5	6	7	8	9	10
2	2	4	6	8	10	12	14	16	18	20
3	3	6	9	12	15	18	21	24	27	30
4	4	8	12	16	20	24	28	32	36	40
5	5	10	15	20	25	30	35	40	45	50
6	6	12	18	24	30	36	42	48	54	60
7	7	14	21	28	35	42	49	56	63	70
8	8	16	24	32	40	48	56	64	72	80
9	9	18	27	36	45	54	63	72	81	90
10	10	20	30	40	50	60	70	80	90	100

TA13

Sieve of Eratosthenes

You probably know the following definitions of prime and composite numbers:

> A **prime number** is a whole number that has exactly two **factors.** The factors are 1 and the number itself. For example, 7 is a prime number because its only factors are 1 and 7. A prime number is divisible by only 1 and itself.

> A **composite number** is a whole number that has more than two factors. For example, 10 is a composite number because it has four factors: 1, 2, 5, and 10. A composite number is divisible by at least three whole numbers.

The number 1 is neither prime nor composite.

For centuries, mathematicians have been interested in prime and composite numbers because they are the building blocks of whole numbers. They have found that every composite number can be written as the product of prime numbers. For example, 18 can be written as 2 * 3 * 3.

Around 300 B.C., the Greek mathematician Euclid (yOO´klid) proved that there is no largest prime number. No matter how large a prime number you find, there will always be larger prime numbers. Since then, people have been searching for more prime numbers. In 1893, a mathematician was able to show that there are more than 50 million prime numbers between the numbers 1 and 1 billion.

The Greek mathematician Eratosthenes (ĕr´ə-tŏs´thə-nēz´), who lived around 200 BCE, devised a simple method for finding prime numbers. His strategy was based on the fact that every **multiple of a number** is divisible by that number. For example, the numbers 2, 4, 6, 8, and 10 are multiples of 2, and each of these numbers is divisible by 2. Here is another way to say it: A whole number is a factor of every one of its multiples. For example, 2 is a factor of 2, 4, 6, 8, and 10. The number 2 has only one other factor, the number 1, so 2 is a prime number. All other multiples of 2 are composite numbers.

Eratosthenes' method is called the **Sieve of Eratosthenes**. The directions for using the sieve to find prime numbers are given on Activity Card 21.

Since the time of Eratosthenes, mathematicians have invented more powerful methods for finding prime numbers. Some methods use formulas. Today, people use computers. The largest prime number known when this book went to press had 17,425,170 digits. If that number were printed in a book with pages the same size as this page, in the same size type, the book would be about up 5,319 pages long.

Sieve of Eratosthenes
(continued)

NAME   DATE   TIME

1	2	3	4	5	6	7	8	9	10
11	12	13	14	15	16	17	18	19	20
21	22	23	24	25	26	27	28	29	30
31	32	33	34	35	36	37	38	39	40
41	42	43	44	45	46	47	48	49	50
51	52	53	54	55	56	57	58	59	60
61	62	63	64	65	66	67	68	69	70
71	72	73	74	75	76	77	78	79	80
81	82	83	84	85	86	87	88	89	90
91	92	93	94	95	96	97	98	99	100

Fact Triangles

NAME   DATE   TIME

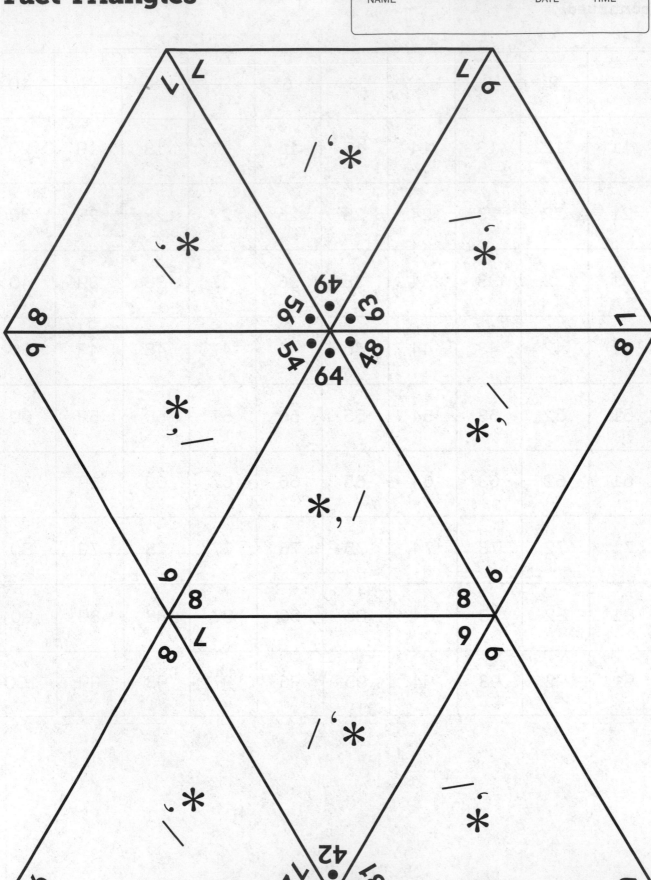

TA16

Blank Comparison Story Cards

TA17

Quadrilateral Cards

NAME   DATE   TIME

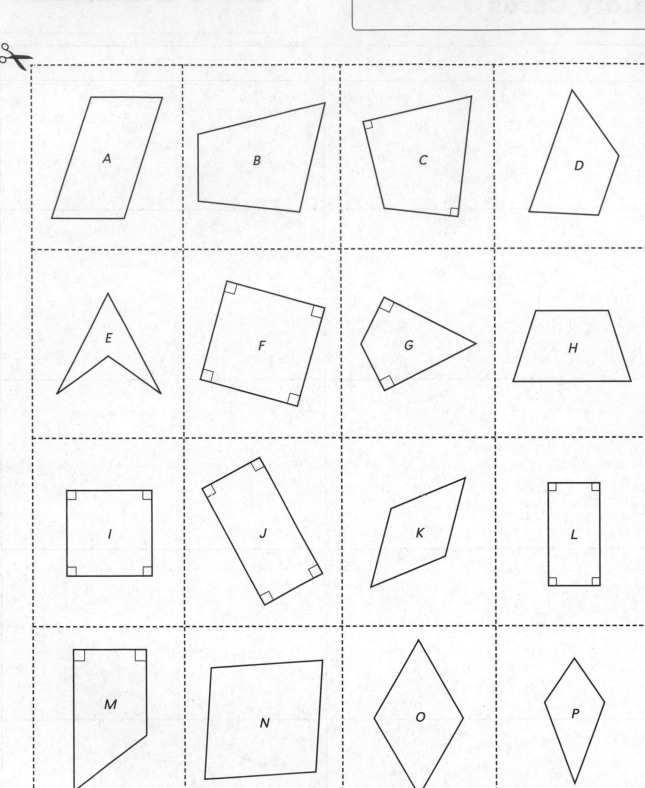

Exit Slip

NAME　　　　　DATE　　TIME

✂ -

Exit Slip

NAME　　　　　DATE　　TIME

"What's My Rule?" Tables

① in
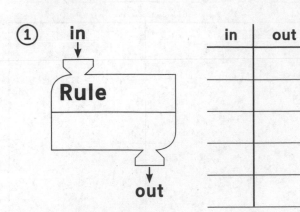
Rule
out

in	out

② in
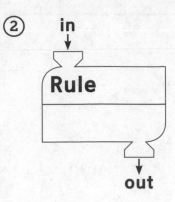
Rule
out

in	out

③ in
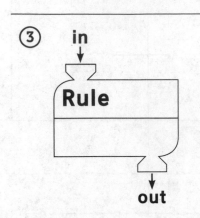
Rule
out

in	out

④ in
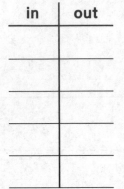
Rule
out

in	out

⑤ in
Rule
out

in	out

⑥ in
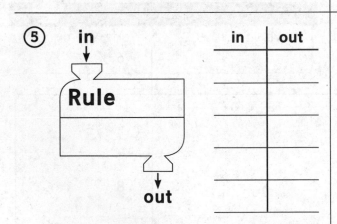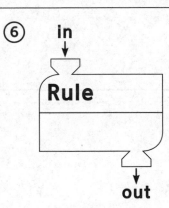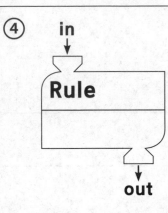
Rule
out

in	out

Number Grid

									0
1	2	3	4	5	6	7	8	9	10
11	12	13	14	15	16	17	18	19	20
21	22	23	24	25	26	27	28	29	30
31	32	33	34	35	36	37	38	39	40
41	42	43	44	45	46	47	48	49	50
51	52	53	54	55	56	57	58	59	60
61	62	63	64	65	66	67	68	69	70
71	72	73	74	75	76	77	78	79	80
81	82	83	84	85	86	87	88	89	90
91	92	93	94	95	96	97	98	99	100
101	102	103	104	105	106	107	108	109	110

Multiplication/Division Diagrams

NAME _____ DATE _____ TIME ____

_____	_____ **per** _____	_____ **in all**

Number model: _____

Answer: _____

Number model with answer: _____

_____	_____ **per** _____	_____ **in all**

Number model: _____

Answer: _____

Number model with answer: _____

_____	_____ **per** _____	**total** _____

Number model: _____

Answer: _____

Number model with answer: _____

TA22

Fraction Circles 1

NAME   DATE   TIME

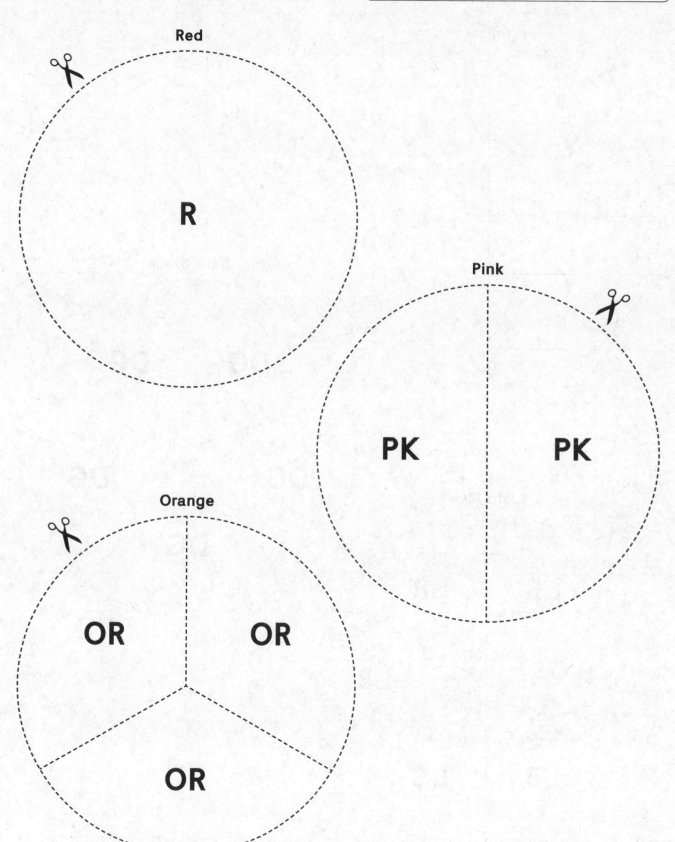

Red

R

Pink

PK   PK

Orange

OR   OR

OR

TA23

Fraction Circles 2

NAME   DATE   TIME

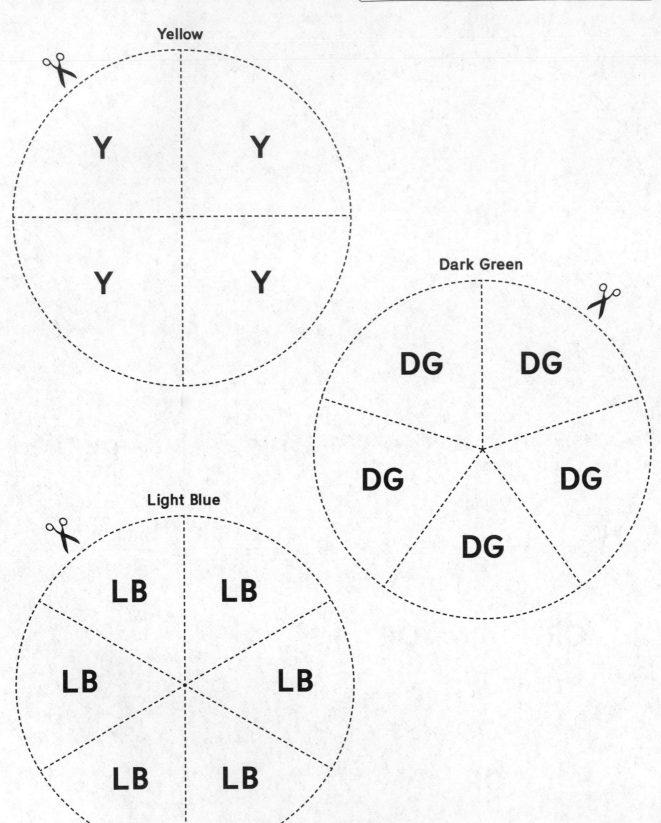

Yellow

Dark Green

Light Blue

TA24

Fraction Circles 3

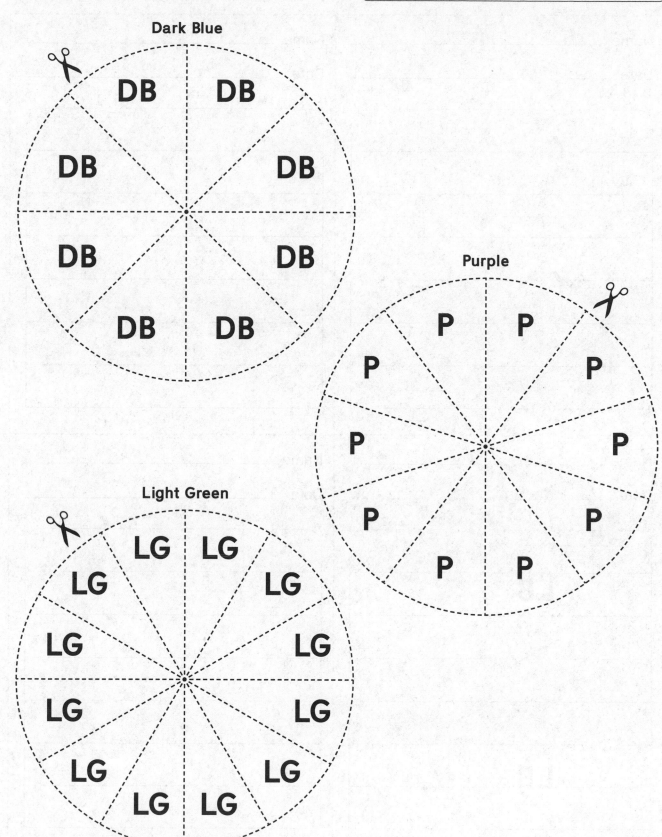

Dark Blue

Purple

Light Green

TA25

Name-Collection Boxes

Name _____

Date _____

Name _____

Date _____

Name _____

Date _____

Name _____

Date _____

A Flat

$1 and $10 Bills

Exploring Decimals

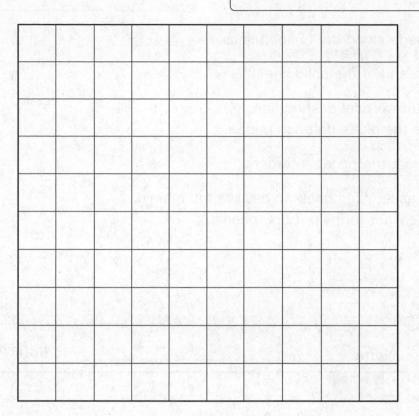

A	B	C	D
_____ hundredths	_____ tenths, _____ hundredths	0._____	
_____ hundredths	_____ tenths, _____ hundredths	0._____	
_____ hundredths	_____ tenths, _____ hundredths	0._____	
_____ hundredths	_____ tenths, _____ hundredths	0._____	
_____ hundredths	_____ tenths, _____ hundredths	0._____	
_____ hundredths	_____ tenths, _____ hundredths	0._____	
_____ hundredths	_____ tenths, _____ hundredths	0._____	

TA29

Decimal Place-Value
Book (Page 1)

① Cut each page along the dashed lines.   - - - - - - - - - - -

   Do NOT cut any of the solid lines!

② Cut along the vertical dashed lines
   to separate the digits on each page.

③ Assemble with the pages in order.

④ Staple the assembled book across the top margin.
   (Ask your teacher for help if you need it.)

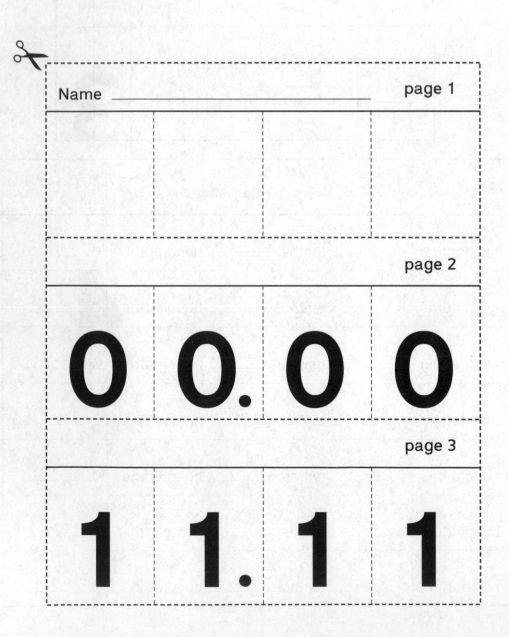

Name _____   page 1

page 2

0   0 . 0   0

page 3

1   1 . 1   1

Decimal Place-Value Book (Page 2)

NAME   DATE   TIME

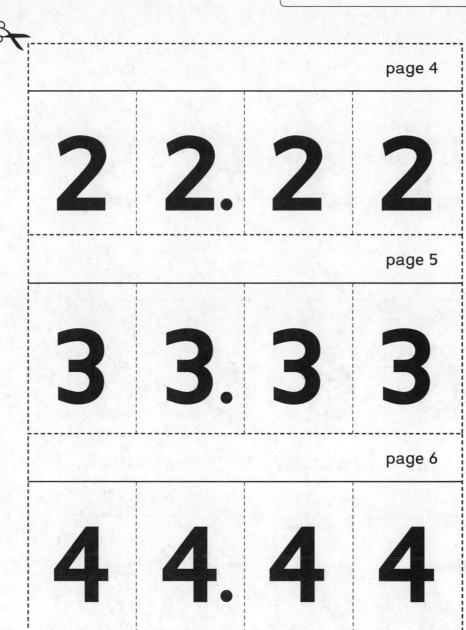

page 4

2 2. 2 2

page 5

3 3. 3 3

page 6

4 4. 4 4

Decimal Place-Value Book (Page 3)

page 7

5   5.   5   5

page 8

6   6.   6   6

page 9

7   7.   7   7

TA32

Decimal Place-Value Book (Page 4)

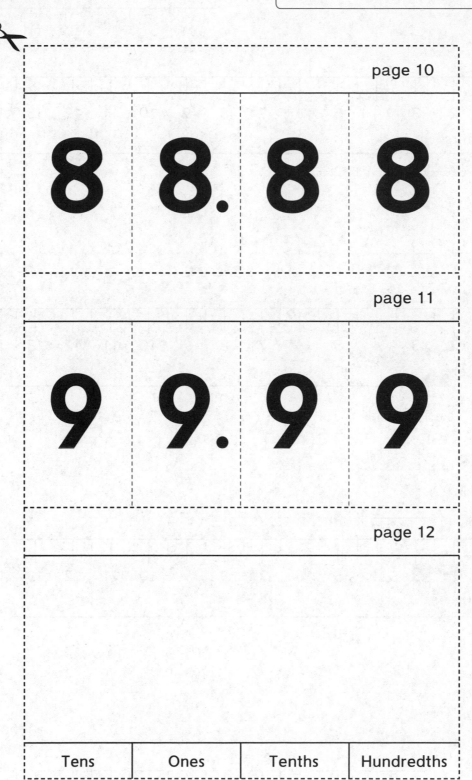

page 10

8	8.	8	8

page 11

9	9.	9	9

page 12

Tens	Ones	Tenths	Hundredths

A cm Ruler

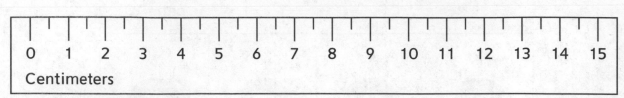

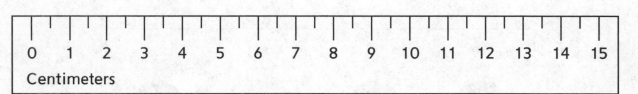

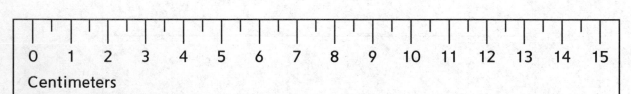

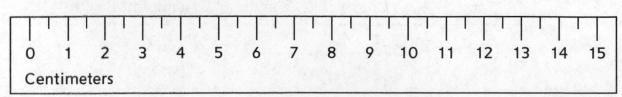

TA34

A cm/mm Ruler

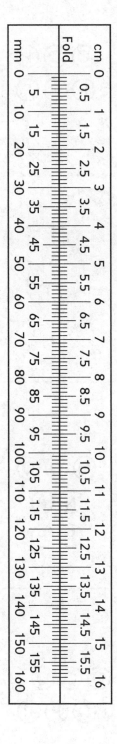

Coin Cards

(N)(N)(P)
(P)(P)(P)

(D)(D)
(N)(P)(P)

(Q)(D)(P)
(P)(P)(P)

(Q)(D)(N)
(P)(P)(P)

(Q)(D)(D)
(P)(P)(P)

(Q)(Q)(D)
(N)(P)(P)

(Q)(Q)(D)
(D)(P)

(Q)(Q)(Q)
(P)(P)(P)

(Q)(Q)(Q)
(N)(P)
(P)(P)

(Q)(Q)(Q)
(D)(N)(P)

(Q)(Q)(Q)
(D)(D)(P)

(Q)(Q)(Q)
(D)(D)(N)

(Q)(Q)(Q)(Q)
(D)(P)(P)

Dollar Bill Cutouts

$10 and $100 Bills

Lattice Multiplication
Computation Grids

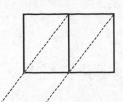
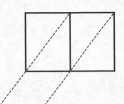
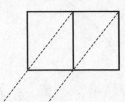

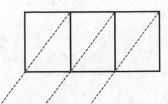
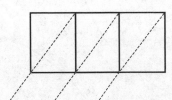
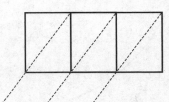

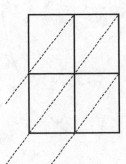
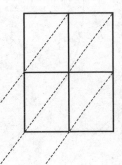
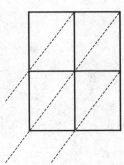

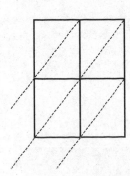
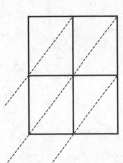
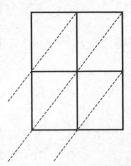

TA39

Fraction Circles

TA40

Tangram Puzzle

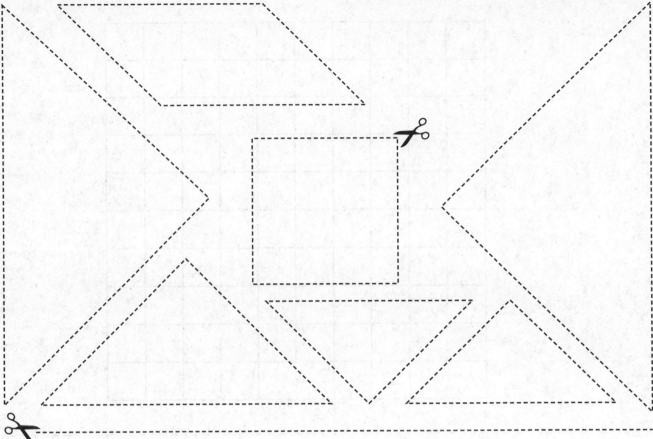

Tenths and Hundredths

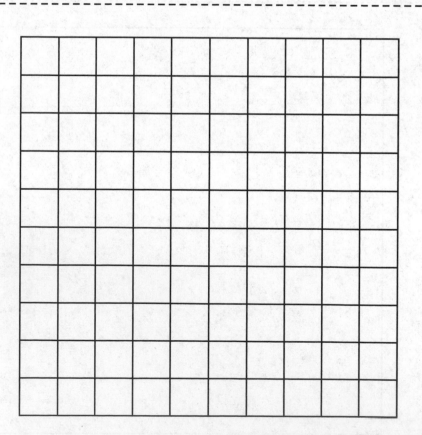

TA42

Frames and Arrows

SRB
62

① **Rule**

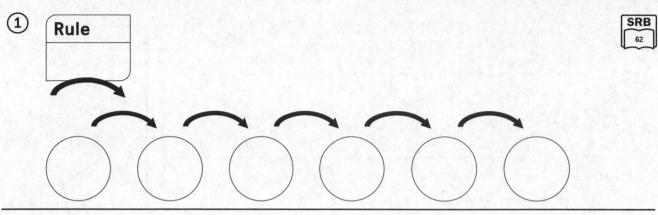

② **Rule**

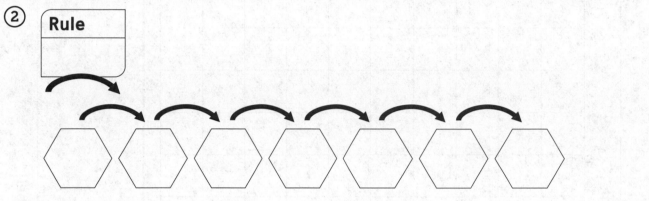

③ **Rule**

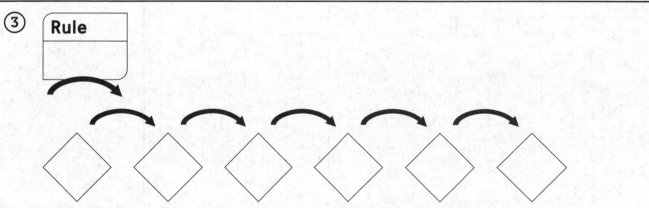

④ **Rule**

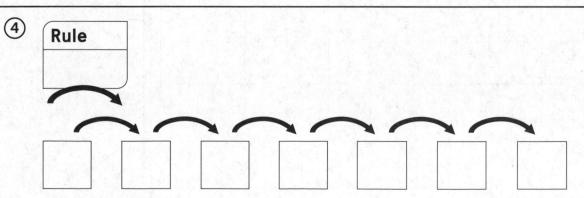

TA43

Making a Line Plot

title

label

TA44

Angle Review Cards

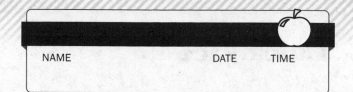
NAME   DATE   TIME

Draw an obtuse angle.	Draw an acute angle.	Draw a right angle.
Draw ray *AB*.	Draw angle *XYZ*.	What type of angle?
What type of angle?	What type of angle?	Name the rays.
Name the vertex.	Draw right angle *JKL*.	Draw acute angle *QRS*.

TA45

Clock Face

① Cut out the clock face, the minute hand, and the hour hand.

② Punch a hole through the center of the clock face and through the X on the minute hand and the hour hand.

③ Fasten the hands to the clock face with a brad.

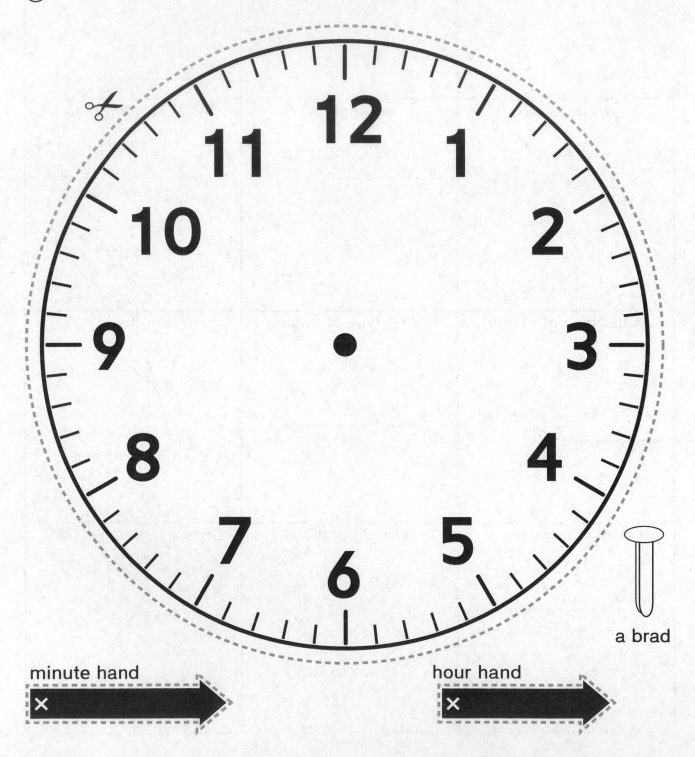

minute hand

hour hand

a brad

TA46

360° Circle

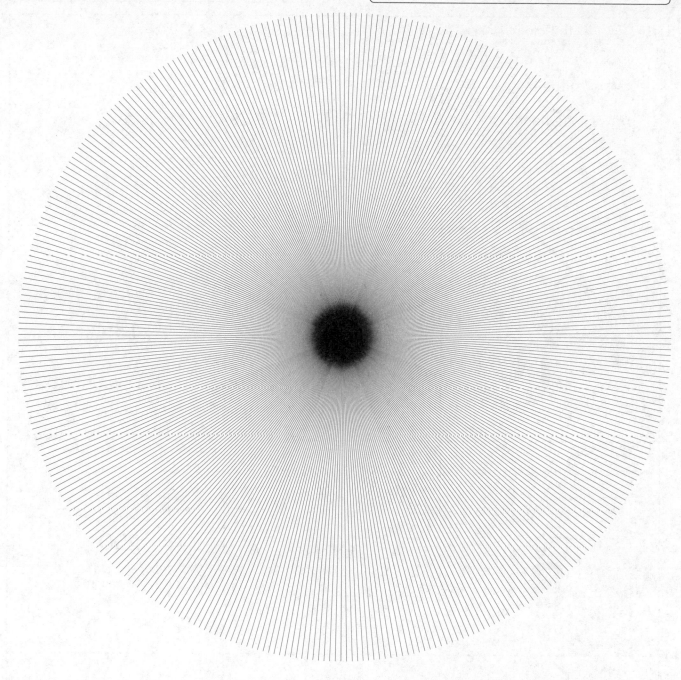

TA47

Angle Measure Cards

5°	10°	30°
45°	100°	135°
170°	200°	250°
300°	315°	355°

Easy Multiples

100 * _____ = _____

50 * _____ = _____

20 * _____ = _____

10 * _____ = _____

5 * _____ = _____

2 * _____ = _____

1 * _____ = _____

100 * _____ = _____

50 * _____ = _____

20 * _____ = _____

10 * _____ = _____

5 * _____ = _____

2 * _____ = _____

1 * _____ = _____

100 * _____ = _____

50 * _____ = _____

20 * _____ = _____

10 * _____ = _____

5 * _____ = _____

2 * _____ = _____

1 * _____ = _____

100 * _____ = _____

50 * _____ = _____

20 * _____ = _____

10 * _____ = _____

5 * _____ = _____

2 * _____ = _____

1 * _____ = _____

TA49

Circles for Angle Measurers

Cut the sheet into four parts along the dashed lines.

Share the circles with the members of your group. Each person will cut out his or her own circle.

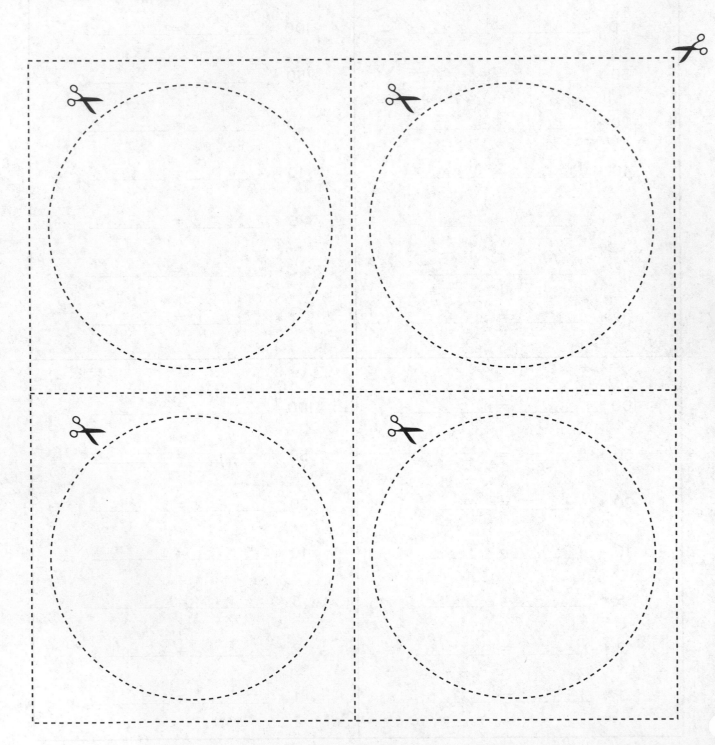

TA50

Circular Geoboard Paper

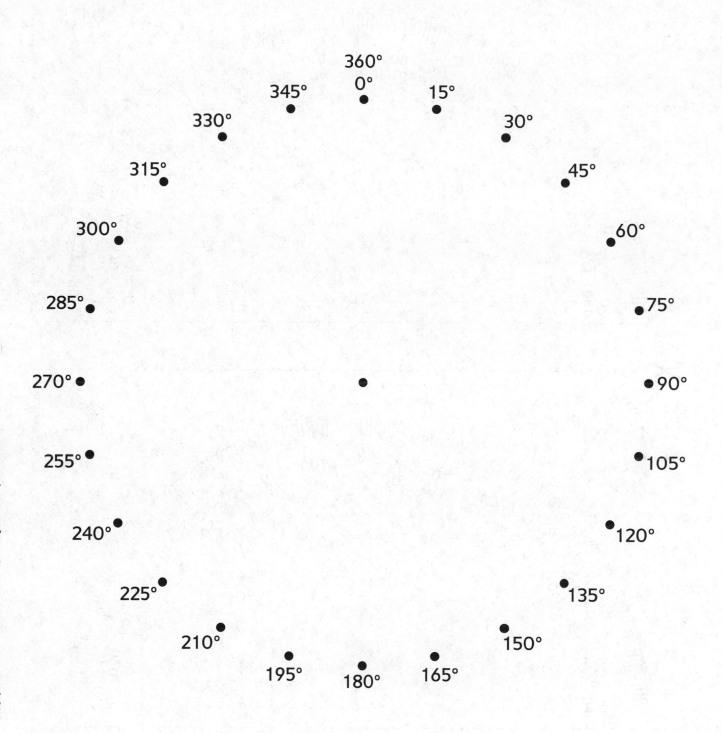

TA51

Half-Circle Protractors (Large)

NAME   DATE   TIME

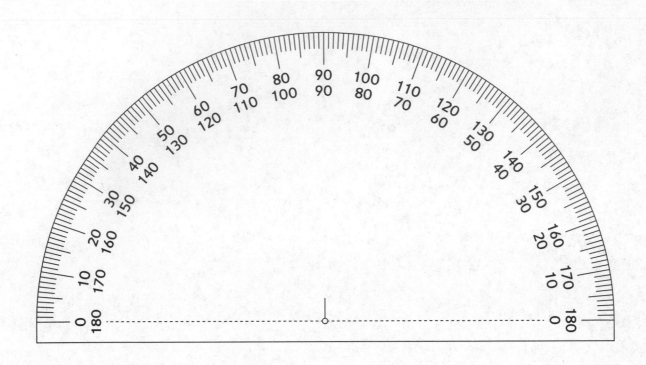

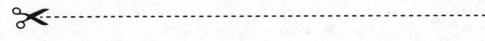

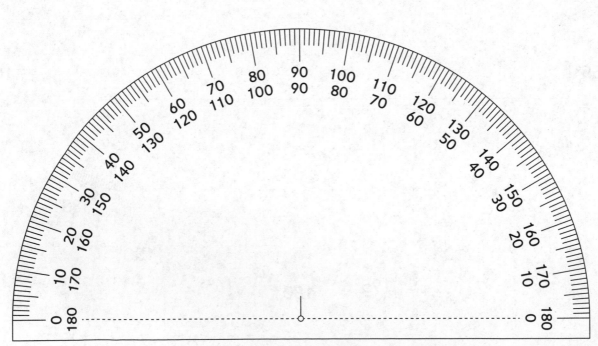

TA52

Half-Circle Protractors
(Small)

NAME   DATE   TIME

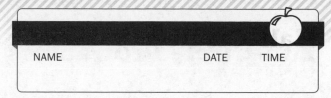

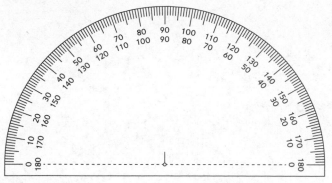

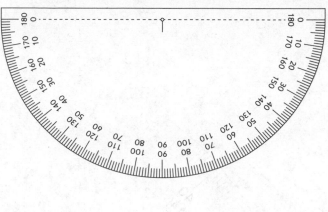

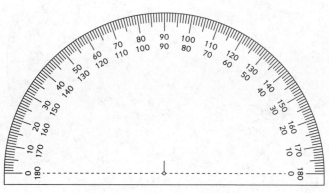

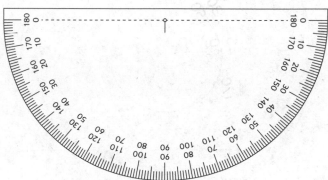

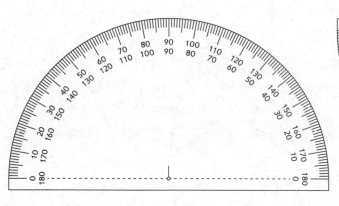

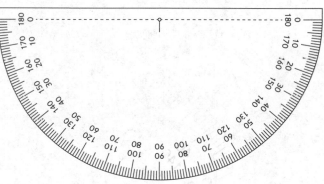

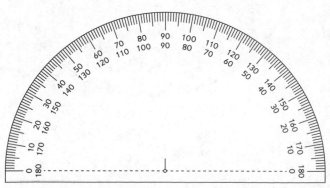

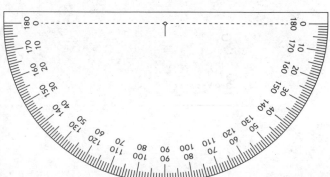

TA53

Full-Circle Protractors

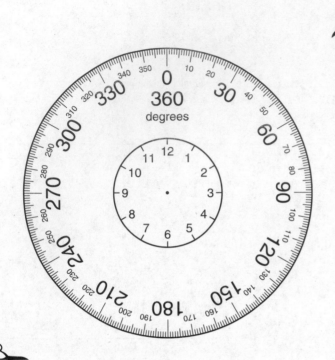

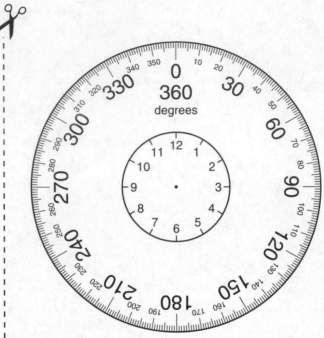

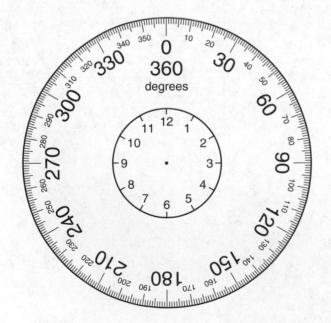

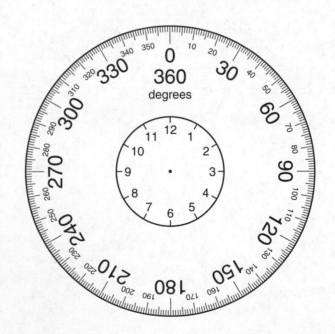

4-Square Graphic Organizer

NAME   DATE   TIME

_____

_____

Word

_____

_____

TA55

Student-Friendly Rubric

NAME   DATE   TIME

Goal:				
Meets Expectations	**Student 1**	**Student 2**	**Student 3**	
Exceeds Expectations				

Stock-Up Sale Poster #1

Lightbulbs
4-Pack   **$1.09**

5 OR MORE SALE | You pay $0.88 per pack.

Extension Cord
$3.25

5 OR MORE SALE | You pay $2.79 per cord.

Tissues
$0.73

5 OR MORE SALE | You pay $0.57 per box.

Transparent Tape
$0.84

5 OR MORE SALE | You pay $0.65 per roll.

Batteries
4-Pack   **$3.59**

5 OR MORE SALE | You pay $2.90 per pack.

Toothpaste
$1.39

5 OR MORE SALE | You pay $1.14 per tube.

Ballpoint Pen
$0.39

5 OR MORE SALE | You pay $0.27 per pen.

Tennis Balls
Can of 3   **$2.59**

5 OR MORE SALE | You pay $1.86 per can.

Paperback Book
$2.99

5 OR MORE SALE | You pay $2.25 per book.

Stock-Up Sale Poster #2

Greeting Cards
Box of 12   **$3.29**

5 OR MORE SALE | You pay $2.63 per box.

Bath Soap
$0.88

5 OR MORE SALE | You pay $0.65 per bar.

Gift Wrapping Paper
$2.35 per roll

5 OR MORE SALE | You pay $1.86 per roll.

Toothbrush
$1.38

5 OR MORE SALE | You pay $1.13 per brush.

Garbage Bags
$3.75

5 OR MORE SALE | You pay $3.18 per box.

Night Light Bulbs
2-Pack   **$0.96**

5 OR MORE SALE | You pay $0.76 per pack.

Glue
$1.15

5 OR MORE SALE | You pay $0.94 per bottle.

Construction Paper
$0.67 per pad

5 OR MORE SALE | You pay $0.54 per pad.

Shoelaces
$1.27 per pair

5 OR MORE SALE | You pay $1.08 per pair.

Inch Rulers

NAME   DATE   TIME

Inches (in.)
0 1 2 3 4 5 6 7 8

Inches (in.)
0 1 2 3 4 5 6 7 8

Inches (in.)
0 1 2 3 4 5 6 7 8

Inches (in.)
0 1 2 3 4 5 6 7 8

Inches (in.)
0 1 2 3 4 5 6 7 8

Math Boxes A

NAME
DATE
TIME

①

②

③

④

Math Boxes B

①

②

③

④

⑤

⑥

TA61

Math Boxes C

(1) Solve.

a. ☐☐ – ☐ = ____

☐☐0 – ☐0 = ____

b. ☐☐ + ☐ = ____

☐☐0 + ☐0 = ____

c. ☐☐ – ☐ = ____

☐☐0 – ☐0 = ____

(2) Use a place-value flip book or chart. In the number

☐☐☐,☐☐☐

the value of ☐ is ☐,000.

a. The ☐ stands for _____.

b. The ☐ stands for _____.

c. The ☐ stands for _____.

d. The ☐ stands for _____.

(3) Fill in the missing numbers.

a. ____, ____, ____, ☐, ☐, ☐

Rule: _____

b. ____, ____, ____, ☐, ☐, ☐

Rule: _____

c. ____, ____, ____, ☐, ☐, ☐

Rule: _____

(4) Add.

a. ☐☐☐

+ ☐☐

‾‾‾‾‾‾

b. ☐☐☐

+ ☐☐

‾‾‾‾‾‾

(5) Write >, <, or = to make each number sentence true.

a. ☐ ____ ☐

b. ☐ ____ ☐

c. ☐ ____ ☐

d. ☐ ____ ☐

(6) Find the area and perimeter of the rectangle below.

☐ ft

☐ ft

Area = ____ square feet

Perimeter = ____ feet

Math Boxes D

① Write a number sentence to estimate the answer to ☐☐ * ☐. Then solve the problem. Show your work.

Estimate:

_____

Answer:

_____

② Multiply.

☐ * ☐ = ☐ _____

☐00 * ☐ = _____

☐,000 * ☐ = _____

③ List the factor pairs for ☐☐.

_____

_____

④ I am the smallest multiple of ☐, ☐, and ☐. What number am I?

_____

⑤ Round each number to the nearest _____.

a. ☐_____ _____

b. ☐_____ _____

c. ☐_____ _____

d. ☐_____ _____

⑥ Write the number.

a. ☐ [1,000s] + ☐ [10s] + ☐ [1s] = _____

b. ☐ [1,000s] + ☐ [10s] + ☐ [1s] = _____

c. ☐ [100,000s] + ☐ [1,000s] + ☐ [100s] = _____

TA63

Math Boxes E

---

(1) Estimate and solve.

[ ] ÷ [ ] = _____

Estimate: _____

Answer: _____

---

(2) Estimate and solve.

[   ]

[ ] ×

[   ]

Estimate: _____

Answer: _____

---

(3) What is the measure of the unknown side?

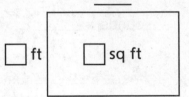

[ ] ft   [ ] sq ft

Equation with unknown:

_____

Answer: _____

---

(4) Divide the number into [   ]ths and label the fractions.

⟵————————————⟶

---

(5) Draw a/an [   ]° angle:

Name the type of angle: _____

---

(6) Complete the table with equivalent names.

Fraction	Decimal

TA64

Math Boxes F

① Write an equivalent fraction or decimal.

Decimal	**Fraction**
a. ☐	_____
b. _____	☐
c. _____	☐
d. ☐	_____

② There are ☐ students who want to share ☐ equally. How many ☐ will each student get?

Answer: _____

Is there a remainder? _____

Is so, what do you need to do with it?

_____

③ Solve.

a. $\dfrac{\boxed{}}{100} + \dfrac{\boxed{}}{10} =$ _____

b. $\dfrac{\boxed{}}{10} + \dfrac{\boxed{}}{100} =$ _____

④ Fill in the blanks:

a. ☐ pounds = _____ ounces

b. ☐ tons = _____ pounds

c. _____ meters = ☐ cm

d. ☐ meter = _____ cm

⑤ Fill in the blanks:

a. ☐ gallons = _____ quarts

b. ☐ pints = _____ cups

c. _____ cups = ☐ fl oz

d. _____ quarts = ☐ pints

⑥

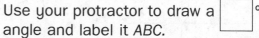

Use your protractor to draw a ☐° angle and label it *ABC*.

Then find the measure of the remaining angle, using an equation.

Equation with unknown:

_____

Answer: _____

Fraction Circles

Red

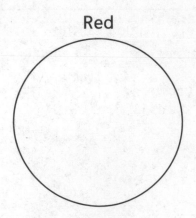

Pink

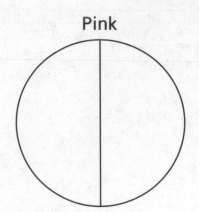

Orange

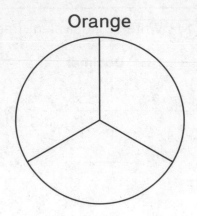

Yellow

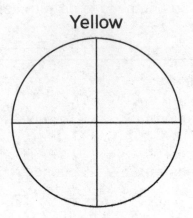

Dark Green

Light Blue

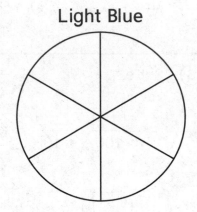

Dark Blue

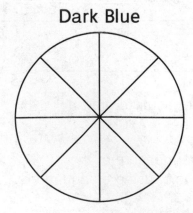

Purple

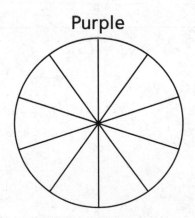

Light Green

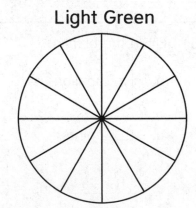

1-Inch Grid Paper

TA67

U.S. Traditional Subtraction Graphic Organizer

NAME DATE TIME

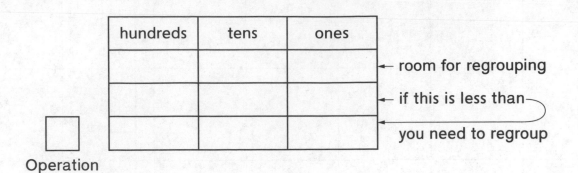

hundreds	tens	ones
			← room for regrouping
			← if this is less than
			you need to regroup

Operation

hundreds	tens	ones
			← room for regrouping
			← if this is less than
			you need to regroup

Operation

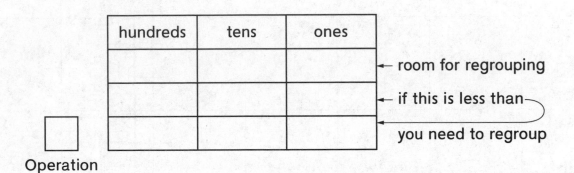

hundreds	tens	ones
			← room for regrouping
			← if this is less than
			you need to regroup

Operation

Fraction/Decimal Equivalency Chart

NAME DATE TIME

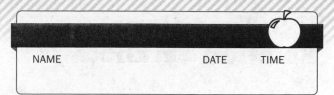

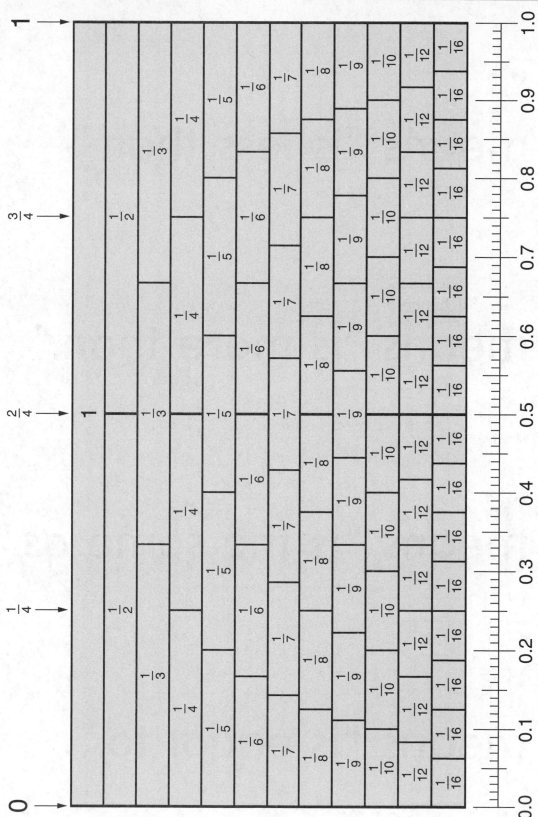

TA69

Relation Symbols

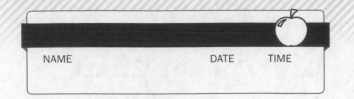

< means "is less than."

> means "is more than."

= means "is the same as."

= means "is equal to."

Shapes and Angles

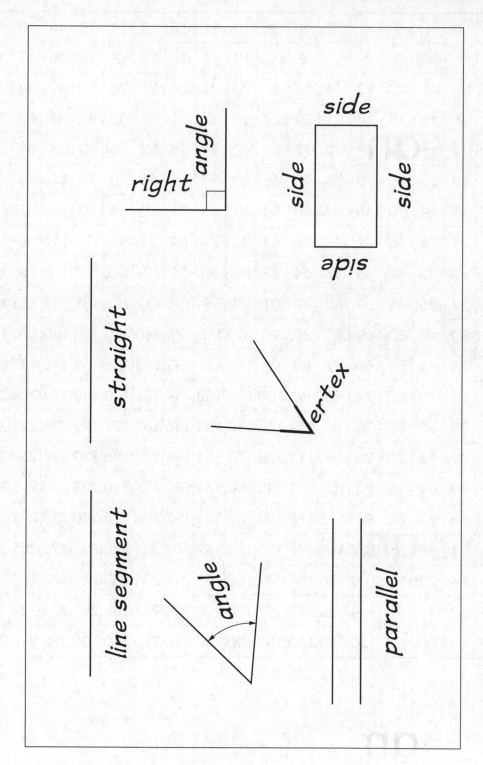

Multiplication Grid (20 by 20)

*	1	2	3	4	5	6	7	8	9	10	11	12	13	14	15	16	17	18	19	20
1	1	2	3	4	5	6	7	8	9	10	11	12	13	14	15	16	17	18	19	20
2	2	4	6	8	10	12	14	16	18	20	22	24	26	28	30	32	34	36	38	40
3	3	6	9	12	15	18	21	24	27	30	33	36	39	42	45	48	51	54	57	60
4	4	8	12	16	20	24	28	32	36	40	44	48	52	56	60	64	68	72	76	80
5	5	10	15	20	25	30	35	40	45	50	55	60	65	70	75	80	85	90	95	100
6	6	12	18	24	30	36	42	48	54	60	66	72	78	84	90	96	102	108	114	120
7	7	14	21	28	35	42	49	56	63	70	77	84	91	98	105	112	119	126	133	140
8	8	16	24	32	40	48	56	64	72	80	88	96	104	112	120	128	136	144	152	160
9	9	18	27	36	45	54	63	72	81	90	99	108	117	126	135	144	153	162	171	180
10	10	20	30	40	50	60	70	80	90	100	110	120	130	140	150	160	170	180	190	200
11	11	22	33	44	55	66	77	88	99	110	121	132	143	154	165	176	187	198	209	220
12	12	24	36	48	60	72	84	96	108	120	132	144	156	168	180	192	204	216	228	240
13	13	26	39	52	65	78	91	104	117	130	143	156	169	182	195	208	221	234	247	260
14	14	28	42	56	70	84	98	112	126	140	154	168	182	196	210	224	238	252	266	280
15	15	30	45	60	75	90	105	120	135	150	165	180	195	210	225	240	255	270	285	300
16	16	32	48	64	80	96	112	128	144	160	176	192	208	224	240	256	272	288	304	320
17	17	34	51	68	85	102	119	136	153	170	187	204	221	238	255	272	289	306	323	340
18	18	36	54	72	90	108	126	144	162	180	198	216	234	252	270	288	306	324	342	360
19	19	38	57	76	95	114	133	152	171	190	209	228	247	266	285	304	323	342	361	380
20	20	40	60	80	100	120	140	160	180	200	220	240	260	280	300	320	340	360	380	400

Organizing Multistep Number Stories

NAME

DATE

TIME

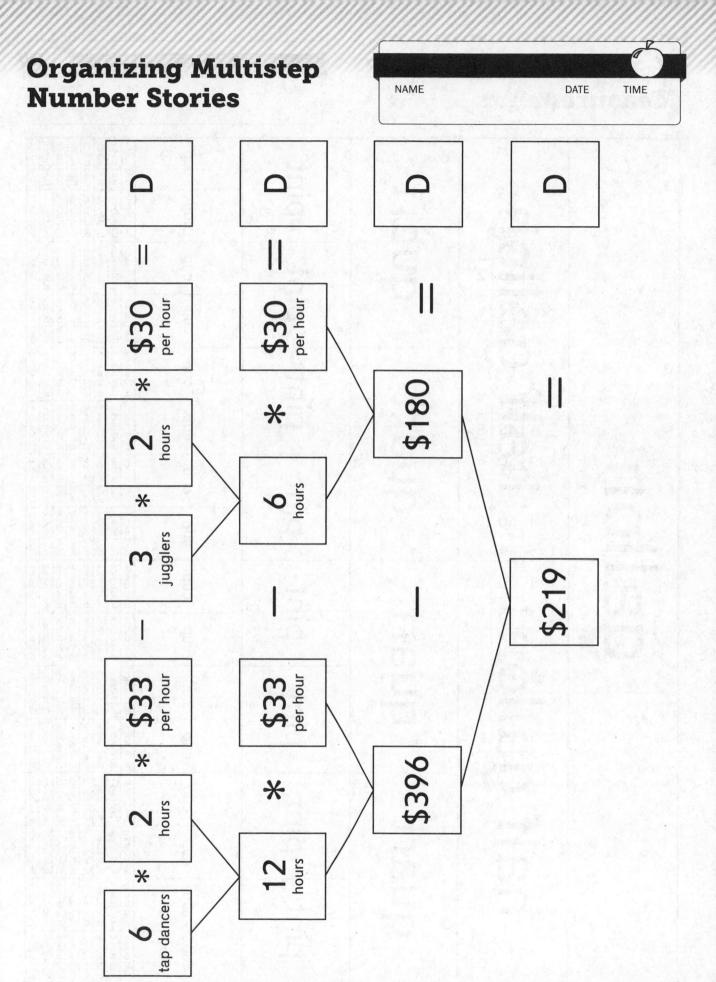

U.S. Traditional Liquid Measurements

gallon	half gallon	quart	pint	cup	fluid ounce
gallon	half gallon	quart	pint	cup	fluid ounce fluid ounce
				cup	fluid ounce fluid ounce
			pint	cup	fluid ounce fluid ounce
				cup	fluid ounce fluid ounce
		quart	pint	cup	fluid ounce fluid ounce
				cup	fluid ounce fluid ounce
			pint	cup	fluid ounce fluid ounce
				cup	fluid ounce fluid ounce
	half gallon	quart	pint	cup	fluid ounce fluid ounce
				cup	fluid ounce fluid ounce
			pint	cup	fluid ounce fluid ounce
				cup	fluid ounce fluid ounce
		quart	pint	cup	fluid ounce fluid ounce
				cup	fluid ounce fluid ounce
			pint	cup	fluid ounce fluid ounce
				cup	fluid ounce fluid ounce

TA74

Guide to Using Easy Multiples +

100 * __4__ = __400__

50 * __4__ = __200__

20 * __4__ = __80__

10 * __4__ = __40__

9 * __4__ = __36__

8 * __4__ = __32__

7 * __4__ = __28__

6 * __4__ = __24__

5 * __4__ = __20__

4 * __4__ = __16__

3 * __4__ = __12__

2 * __4__ = __8__

1 * __4__ = __4__

1. Record the problem.

    4)‾112‾

2. Circle the divisor.

    ④)‾112‾

3. Fill out the Easy Multiples + chart for the divisor.

4. Use the chart to help you find a series of "at least/not more than" easy multiples of the divisor. Ask are there at least 20 [4s] in 112?

5. Write the first partial quotient on the right.

    4)‾112‾ | 20
      80 |

6. Subtract 80 from 112 to see what is left to divide.

7. Go back to Steps 5 and 6, and repeat until the subtraction leaves a number less than the divisor.

    4)‾112‾ | 20
      80 |
      32 | 8
      32 |
       0 |

8. Add the partial quotients to get the quotient (answer).

    4)‾112‾ | 20
      80 |
      32 |
      32 | 8
       0 | 28

TA75

Easy Multiples +

NAME   DATE   TIME

100 * _____	=	_____
50 * _____	=	_____
20 * _____	=	_____
10 * _____	=	_____
9 * _____	=	_____
8 * _____	=	_____
7 * _____	=	_____
6 * _____	=	_____
5 * _____	=	_____
4 * _____	=	_____
3 * _____	=	_____
2 * _____	=	_____
1 * _____	=	_____

100 * _____	=	_____
50 * _____	=	_____
20 * _____	=	_____
10 * _____	=	_____
9 * _____	=	_____
8 * _____	=	_____
7 * _____	=	_____
6 * _____	=	_____
5 * _____	=	_____
4 * _____	=	_____
3 * _____	=	_____
2 * _____	=	_____
1 * _____	=	_____

Fraction Number Lines

NAME   DATE   TIME

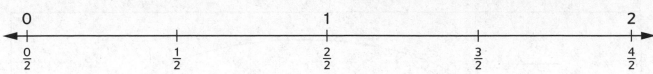

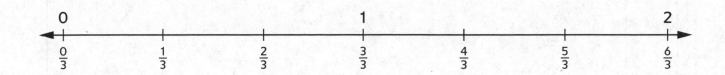

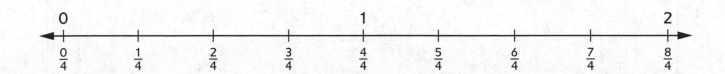

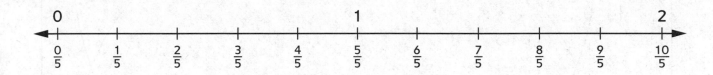

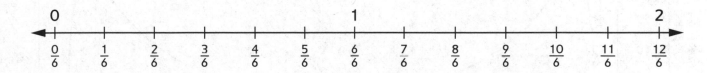

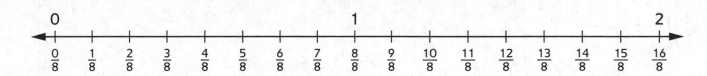

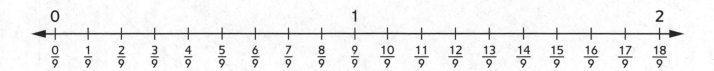

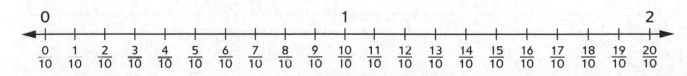

TA77

Venn Diagram

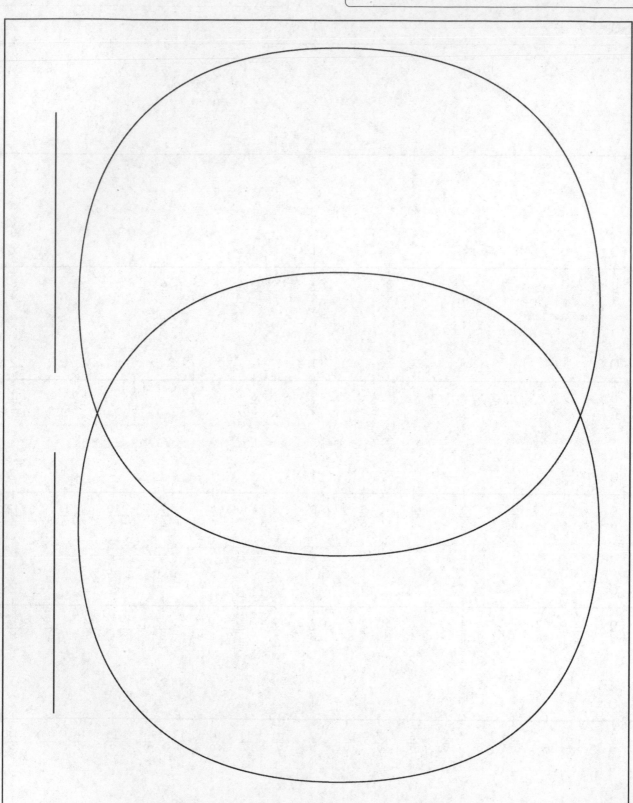

TA78

Parts-and-Total Diagram

Total

Part

Part

TA79

Function Machine

in

↓

Rule

↓

out

in	out

Frames and Arrows

① **Rule**

② **Rule**

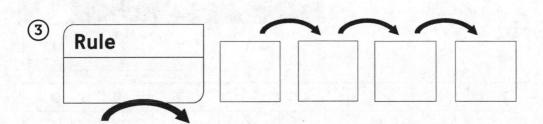

③ **Rule**

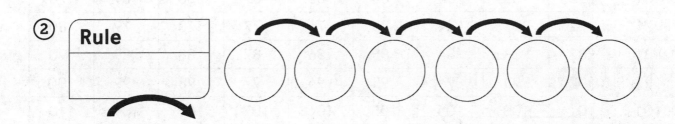

④ **Rule**

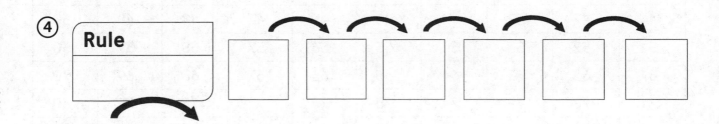

TA81

Number Grid

NAME							DATE		TIME

−9	−8	−7	−6	−5	−4	−3	−2	−1	0
1	2	3	4	5	6	7	8	9	10
11	12	13	14	15	16	17	18	19	20
21	22	23	24	25	26	27	28	29	30
31	32	3?	34	35	36	37	38	39	40
41	42	43	44	45	46	47	48	49	50
51	52	53	54	55	56	57	58	59	60
61	62	63	64	65	66	67	68	69	70
71	72	73	74	75	76	77	78	79	80
81	82	83	84	85	86	87	88	89	90
91	92	93	94	95	96	97	98	99	100
101	102	103	104	105	106	107	108	109	110

Number Grid

NAME							DATE		TIME

−9	−8	−7	−6	−5	−4	−3	−2	−1	0
1	2	3	4	5	6	7	8	9	10
11	12	13	14	15	16	17	18	19	20
21	22	23	24	25	26	27	28	29	30
31	32	3?	34	35	36	37	38	39	40
41	42	43	44	45	46	47	48	49	50
51	52	53	54	55	56	57	58	59	60
61	62	63	64	65	66	67	68	69	70
71	72	73	74	75	76	77	78	79	80
81	82	83	84	85	86	87	88	89	90
91	92	93	94	95	96	97	98	99	100
101	102	103	104	105	106	107	108	109	110

Number Lines

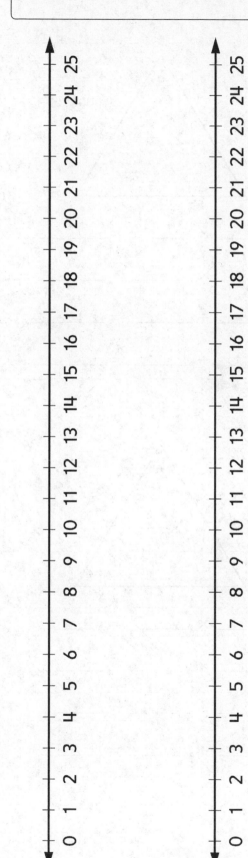

0 1 2 3 4 5 6 7 8 9 10 11 12 13 14 15 16 17 18 19 20 21 22 23 24 25

0 1 2 3 4 5 6 7 8 9 10 11 12 13 14 15 16 17 18 19 20 21 22 23 24 25

0 1 2 3 4 5 6 7 8 9 10 11 12 13 14 15 16 17 18 19 20 21 22 23 24 25

0 1 2 3 4 5 6 7 8 9 10 11 12 13 14 15 16 17 18 19 20 21 22 23 24 25

TA83

Fact Triangles

NAME   DATE   TIME

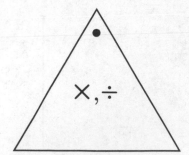

×,÷

_____ × _____ = _____

_____ × _____ = _____

_____ ÷ _____ = _____

_____ ÷ _____ = _____

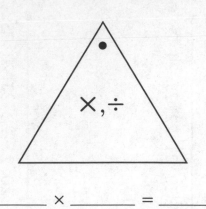

×,÷

_____ × _____ = _____

_____ × _____ = _____

_____ ÷ _____ = _____

_____ ÷ _____ = _____

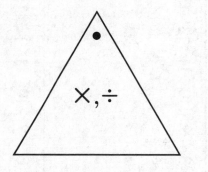

×,÷

_____ × _____ = _____

_____ × _____ = _____

_____ ÷ _____ = _____

_____ ÷ _____ = _____

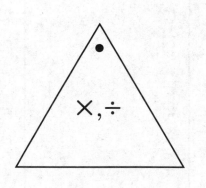

×,÷

_____ × _____ = _____

_____ × _____ = _____

_____ ÷ _____ = _____

_____ ÷ _____ = _____

Game Masters

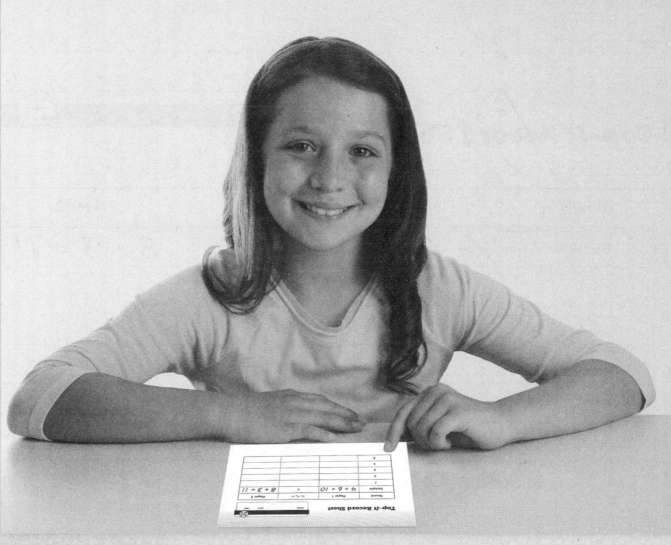

Top-It Record Sheet

Round	Player 1	>, <, =	Player 2
Sample	4 + 6 = 10	<	8 + 3 = 11
1			
2			
3			
4			
5			

Top-It Record Sheet

Round	Player 1	>, <, =	Player 2
Sample	4 + 6 = 10	<	8 + 3 = 11
1			
2			
3			
4			
5			

Number Top-It Mat

Hundred-Thousands

Ten-Thousands

Thousands

G3

Number Top-It Mat

(continued)

NAME   DATE   TIME

Ones

Tens

Hundreds

Do not cut. Paste or tape to *Math Masters*, page G3.

Spin-and-Round
Record Sheet

NAME   DATE   TIME

Circle the place-value digit in the starting number to indicate the spinner directions. For example, circle the 6 in the starting number 462,017 if you are rounding to the nearest ten-thousand.

SRB
273

Player 1 _____   **Score**   **Player 2** _____   **Score**

Starting __ __ __, __ __ __   Starting __ __ __, __ __ __

Rounded __ __ __, __ __ __   ____   Rounded __ __ __, __ __ __   ____

Starting __ __ __, __ __ __   Starting __ __ __, __ __ __

Rounded __ __ __, __ __ __   ____   Rounded __ __ __, __ __ __   ____

Starting __ __ __, __ __ __   Starting __ __ __, __ __ __

Rounded __ __ __, __ __ __   ____   Rounded __ __ __, __ __ __   ____

Starting __ __ __, __ __ __   Starting __ __ __, __ __ __

Rounded __ __ __, __ __ __   ____   Rounded __ __ __, __ __ __   ____

Starting __ __ __, __ __ __   Starting __ __ __, __ __ __

Rounded __ __ __, __ __ __   ____   Rounded __ __ __, __ __ __   ____

Starting __ __ __, __ __ __   Starting __ __ __, __ __ __

Rounded __ __ __, __ __ __   ____   Rounded __ __ __, __ __ __   ____

Starting __ __ __, __ __ __   Starting __ __ __, __ __ __

Rounded __ __ __, __ __ __   Rounded __ __ __, __ __ __

             Total Score ____   **Total Score** ____

Spin-and-Round
Spinner

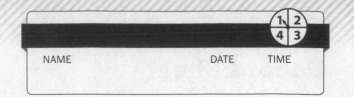
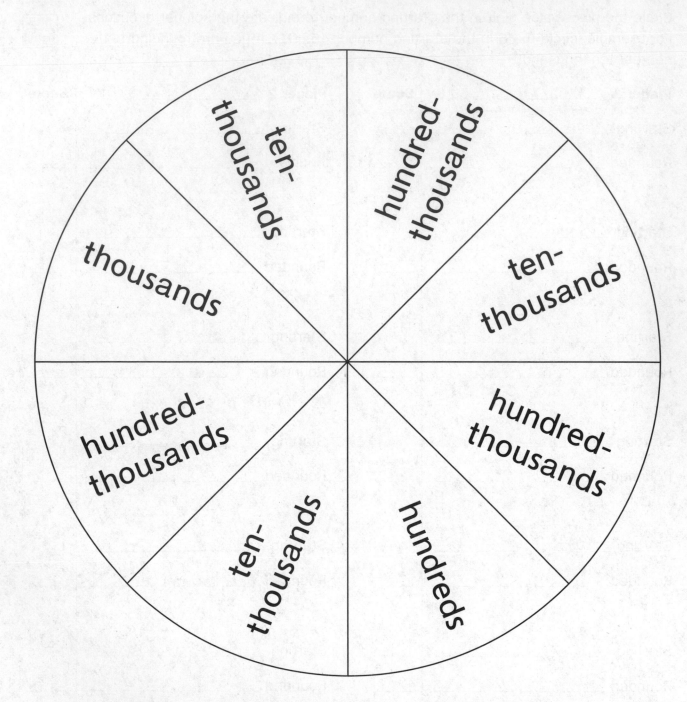

Fishing for Digits
Record Sheet

	Beginning Number	X						
1	New Number							
	New Number							
2	New Number							
	New Number							
3	New Number							
	New Number							
4	New Number							
	New Number							
5	New Number							
	Final Number							

Fishing for Digits
Record Sheet

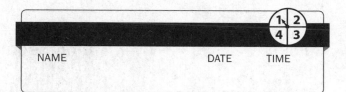

	Beginning Number	X						
1	New Number							
	New Number							
2	New Number							
	New Number							
3	New Number							
	New Number							
4	New Number							
	New Number							
5	New Number							
	Final Number							

Geometry Concentration (Part 1)

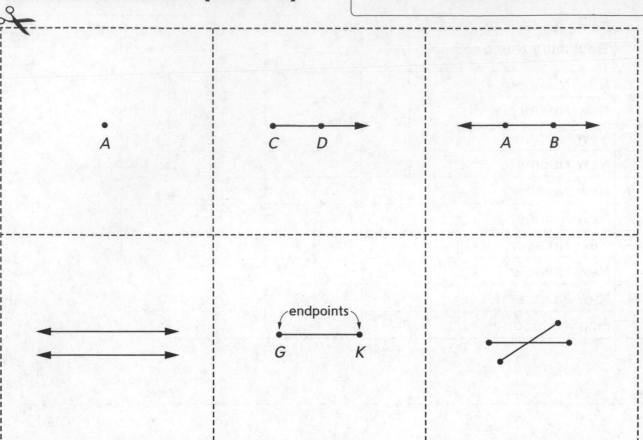

Point	**Line Segment**	**Ray**
A location in space named with a letter	A straight path between 2 points called its endpoints	A straight path that goes on forever in one direction from an endpoint
Parallel Lines	**Line**	**Intersecting Line Segments**
Lines that never cross or meet and are the same distance apart everywhere	A straight path that goes on forever in both directions	Line segments that cross or meet

Geometry Concentration (Part 2)

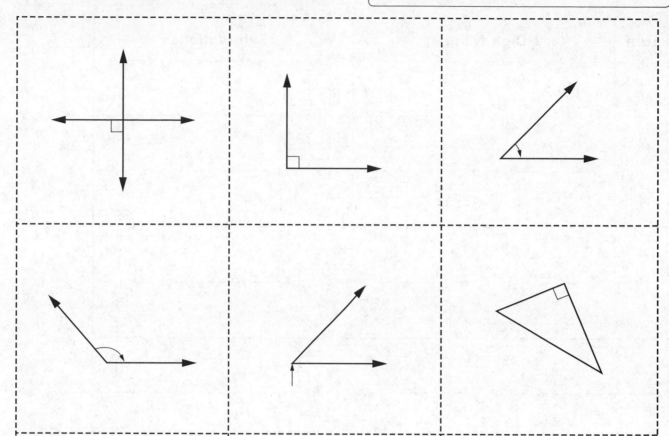

Perpendicular Lines	**Right Angle**	**Acute Angle**
Lines that intersect to form right angles	An angle that measures 90 degrees	An angle that measures less than 90 degrees
Obtuse Angle	**Vertex**	**Right Triangle**
An angle that measures more than 90 degrees and less than 180 degrees	Endpoint where rays or segments meet	3-sided shape with one right angle

G9

Subtraction Target Practice Record Sheet

Turn	2-Digit Number	Subtraction
1		
2		
3		
4		
5		

SRB
274

✂ -

Subtraction Target Practice Record Sheet

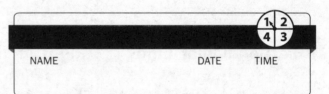

Turn	2-Digit Number	Subtraction
1		
2		
3		
4		
5		

SRB
274

Rugs and Fences Cards

A	*A*	*A*	*A*
Find the area.	**Find the area.**	**Find the area.**	**Find the area.**
P	*P*	*P*	*P*
Find the perimeter.	**Find the perimeter.**	**Find the perimeter.**	**Find the perimeter.**
A or *P*	*A* or *P*	*A* or *P*	*A* or *P*
Partner's Choice	**Partner's Choice**	**Partner's Choice**	**Partner's Choice**
A or *P*	*A* or *P*	*A* or *P*	*A* or *P*
Player's Choice	**Player's Choice**	**Player's Choice**	**Player's Choice**

G11

Rugs and Fences
Rectangle Cards

NAME   DATE   TIME

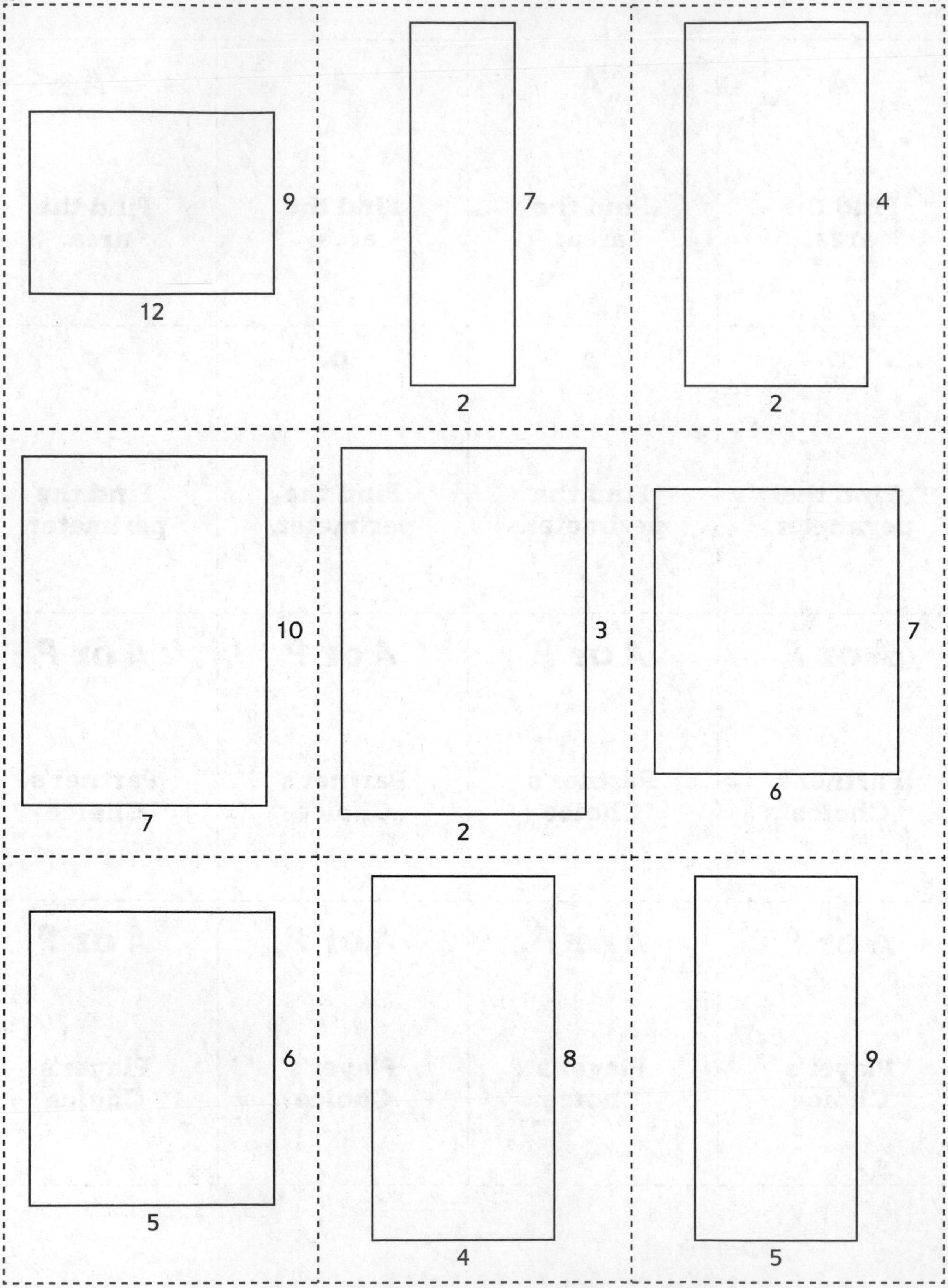

9

12

7

2

4

2

10

7

3

2

7

6

6

5

8

4

9

5

Rugs and Fences
Rectangle Cards (continued)

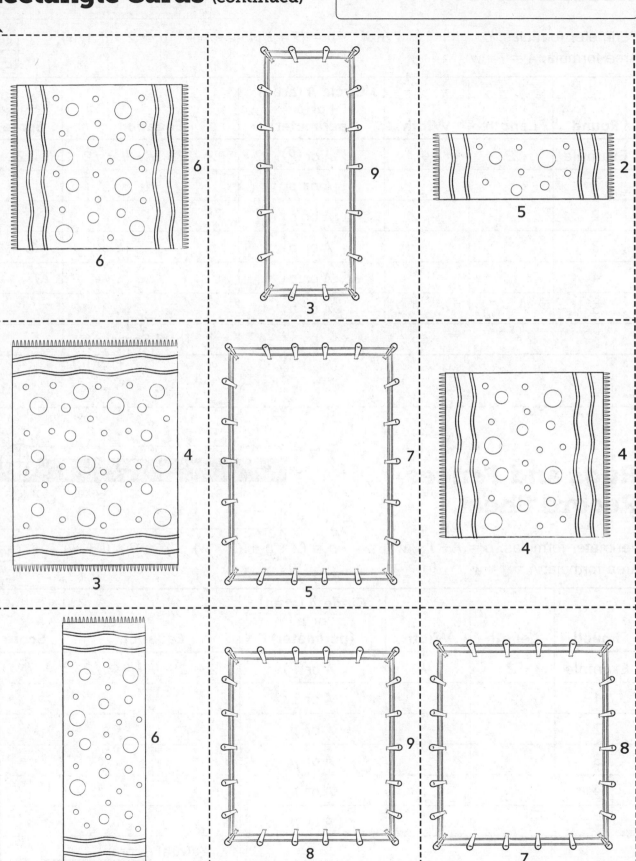

Rugs and Fences
Record Sheet

NAME   DATE   TIME

Perimeter formulas: $p = l + l + w + w$ $p = (2 * l) + (2 * w)$ $p = 2 * (l + w)$

Area formula: $A = l * w$

Round	Length	Width	Circle A (area) or p (perimeter)	Equation	Score
Example	2	4	A or ⓟ	2 + 2 + 4 + 4 = 12	12
1			A or p		
2			A or p		
3			A or p		
4			A or p		
5			A or p		
				Total Score	

✂ -

Rugs and Fences
Record Sheet

NAME   DATE   TIME

Perimeter formulas: $p = l + l + w + w$ $p = (2 * l) + (2 * w)$ $p = 2 * (l + w)$

Area formula: $A = l * w$

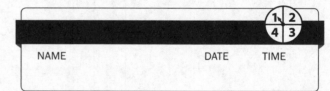

Round	Length	Width	Circle A (area) or p (perimeter)	Equation	Score
Example	2	4	A or ⓟ	2 + 2 + 4 + 4 = 12	12
1			A or p		
2			A or p		
3			A or p		
4			A or p		
5			A or p		
				Total Score	

Factor Captor
Game Mat—Grid 1

NAME   DATE   TIME

1	2	2	2	2	2
2	3	3	3	3	3
3	4	4	4	4	5
5	5	5	6	6	7
7	8	8	9	9	10
10	11	12	13	14	15
16	18	20	21	22	24
25	26	27	28	30	32

Factor Captor
Game Mat—Grid 2

1	2	2	2	2	2	3
3	3	3	3	4	4	4
4	5	5	5	5	6	6
6	7	7	8	8	9	9
10	10	11	12	13	14	15
16	17	18	19	20	21	22
23	24	25	26	27	28	30
32	33	34	35	36	38	39
40	42	44	45	46	48	49
50	51	52	54	55	56	60

Factor Captor
1–110 Grid

1	2	3	4	5	6	7	8	9	10
11	12	13	14	15	16	17	18	19	20
21	22	23	24	25	26	27	28	29	30
31	32	33	34	35	36	37	38	39	40
41	42	43	44	45	46	47	48	49	50
51	52	53	54	55	56	57	58	59	60
61	62	63	64	65	66	67	68	69	70
71	72	73	74	75	76	77	78	79	80
81	82	83	84	85	86	87	88	89	90
91	92	93	94	95	96	97	98	99	100
101	102	103	104	105	106	107	108	109	110

Multiples Bingo
Board

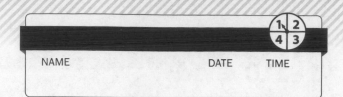

NAME   DATE   TIME

✂ -

Multiples Bingo
Board

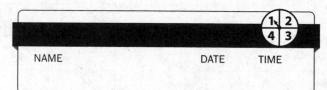

NAME   DATE   TIME

Factor Bingo Game Mat

NAME   DATE TIME

Write any of the numbers 2 through 90 on the grid above.

You may use a number only once.

To help you keep track of the numbers you use, circle them in the list.

2	3	4	5	6	7	8	9	10
11	12	13	14	15	16	17	18	19	20
21	22	23	24	25	26	27	28	29	30
31	32	33	34	35	36	37	38	39	40
41	42	43	44	45	46	47	48	49	50
51	52	53	54	55	56	57	58	59	60
61	62	63	64	65	66	67	68	69	70
71	72	73	74	75	76	77	78	79	80
81	82	83	84	85	86	87	88	89	90

G19

How Much More?
Story Cards

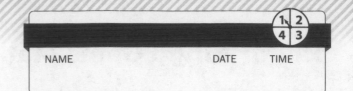
Imani has ____ seashells in her collection. Maggie has ____ more seashells than Imani. How many seashells does Maggie have?	Jeremiah practiced the guitar for ____ hours last month. Alexander practiced the guitar for ____ hours more than Jeremiah. How many hours did Alexander practice?
Joe's dog weighs ____ pounds. Tom's dog weighs ____ pounds more than Joe's dog. How many pounds does Tom's dog weigh?	A piggy bank has nickels and dimes in it. It has ____ nickels. The piggy bank has ____ times as many dimes as nickels. How many dimes are in the piggy bank?
A black rock weighs ____ pounds. A brown rock is ____ times as heavy. How heavy is the brown rock?	Olivia has ____ beads. Betsey has ____ times as many beads. How many beads does Betsey have?
Jackson's classroom has ____ pencils. Zari's classroom has ____ times as many pencils as Jackson's classroom. How many pencils are in Zari's classroom?	Ben planted ____ seeds in the school garden. Wes planted ____ times as many seeds. How many seeds did Wes plant?
Noah saved $____ last week. This week he saved ____ times as much money. How much money did Noah save this week?	Tatum has ____ stuffed animals. Jordan has ____ times as many stuffed animals as Tatum. How many stuffed animals does Jordan have?

How Much More?
Record Sheet

NAME DATE TIME

Additive (A) or Multiplicative (M)	Quantity (first roll)	Number more than OR Number of times as much as (second roll)	Quantity (answer)	Comparison equation
M	9	6	54	9 * 6 = 54
Total Points:				

✂ -

How Much More?
Record Sheet

NAME DATE TIME

Additive (A) or Multiplicative (M)	Quantity (first roll)	Number more than OR Number of times as much as (second roll)	Quantity (answer)	Comparison equation
M	9	6	54	9 * 6 = 54
Total Points:				

Polygon Capture Pieces

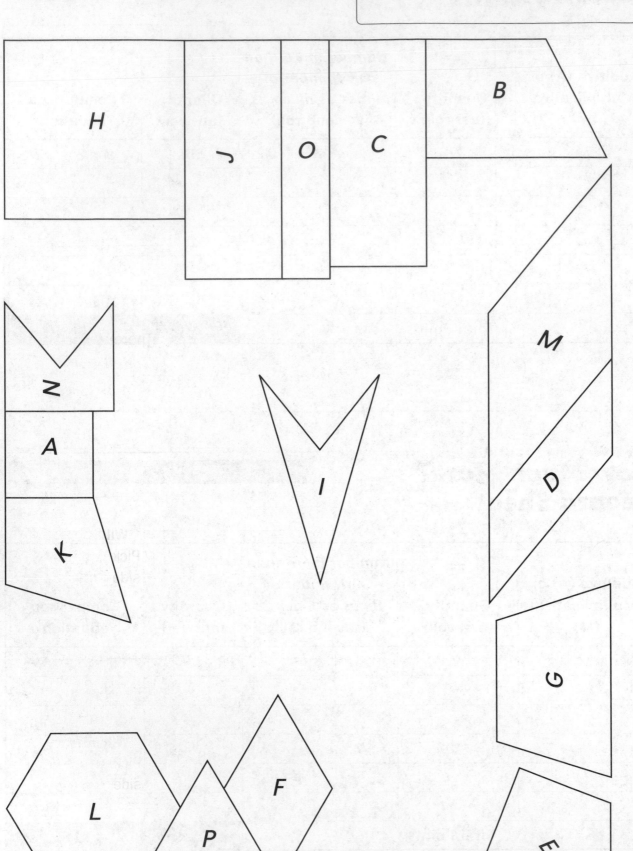

Polygon Capture
Property Cards

There is only one right angle.	There are one or more right angles.	All angles are right angles.	There are no right angles.
There is at least one acute angle.	At least one angle is more than 90°.	All angles are right angles.	There are no right angles.
All opposite sides are parallel.	Only one pair of sides is parallel.	There are no parallel sides.	**Wild Card:** Pick your own side property.
At least two sides are perpendicular.	There are four perpendicular sides.	All the sides are the same length.	**Wild Card:** Pick your own side property.

G23

Polygon Capture
Property Cards (continued)

Angles	Angles	Angles	Angles
Angles	Angles	Angles	Angles
Sides	Sides	Sides	Sides
Sides	Sides	Sides	Sides

Polygon Capture
Record Sheet

NAME   DATE   TIME

Round	Property (Properties)	List Polygon(s) Captured	Number of Polygons Captured
1			
2			
3			
4			
5			
		TOTAL	

Geometry
Concentration, Part 3

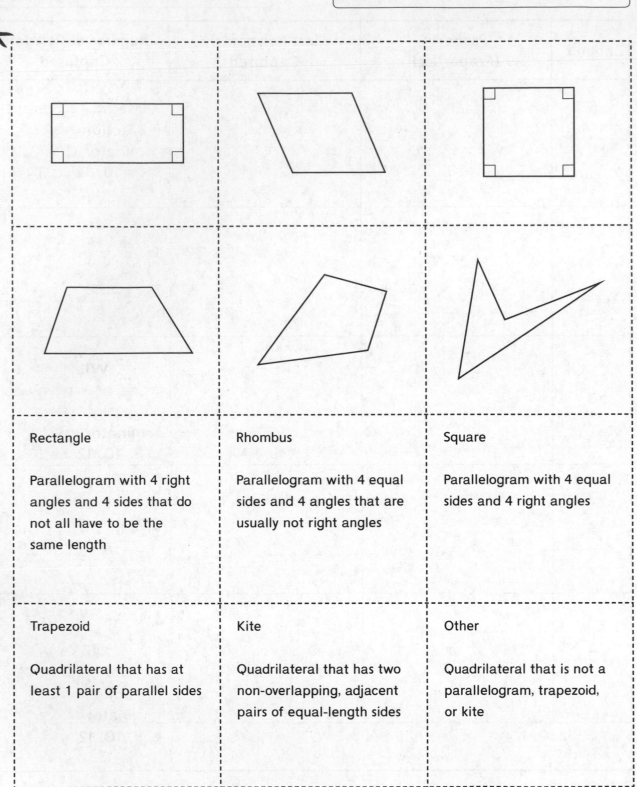

Rectangle

Parallelogram with 4 right angles and 4 sides that do not all have to be the same length

Rhombus

Parallelogram with 4 equal sides and 4 angles that are usually not right angles

Square

Parallelogram with 4 equal sides and 4 right angles

Trapezoid

Quadrilateral that has at least 1 pair of parallel sides

Kite

Quadrilateral that has two non-overlapping, adjacent pairs of equal-length sides

Other

Quadrilateral that is not a parallelogram, trapezoid, or kite

WILD Cards

WILD WILD	WILD WILD	WILD WILD
WILD Name an equivalent fraction with a denominator of 2, 3, 4, 5, 6, 8, 10, 12, or 100.	**WILD** Name an equivalent fraction with a denominator of 2, 3, 4, 5, 6, 8, 10, 12, or 100.	**WILD** Name an equivalent fraction with a denominator of 2, 3, 4, 5, 6, 8, 10, 12, or 100.
WILD WILD	WILD WILD	WILD WILD
WILD Name an equivalent fraction with a denominator of 2, 3, 4, 5, 6, 8, 10, 12, or 100.	**WILD** Name an equivalent fraction with a denominator of 2, 3, 4, 5, 6, 8, 10, 12, or 100.	**WILD** Name an equivalent fraction with a denominator of 2, 3, 4, 5, 6, 8, 10, 12, or 100.
WILD WILD	WILD WILD	WILD WILD
WILD Name an equivalent fraction with a denominator of 2, 3, 4, 5, 6, 8, 10, 12, or 100.	**WILD** Name an equivalent fraction with a denominator of 2, 3, 4, 5, 6, 8, 10, 12, or 100.	**WILD** Name an equivalent fraction with a denominator of 2, 3, 4, 5, 6, 8, 10, 12, or 100.

Fraction Match
Record Sheet

NAME | DATE | TIME

Select two fractions that match.

_____   _____

Explain how you know the fractions are equivalent.

_____

_____

_____

- -

Fraction Match
Record Sheet

NAME | DATE | TIME

Select two fractions that match.

_____   _____

Explain how you know the fractions are equivalent.

_____

_____

_____

Base-10
Decimal Exchange

$1.00
1

Dollars Ones Flats

$0.10
0.1

Dimes Tenths Longs

$0.01
0.01

Pennies Hundredths Cubes

Decimal Top-It Mat

Ones

Tenths

Hundredths

G30

Factor Captor
1–110 Grid

1	2	3	4	5	6	7	8	9	10
11	12	13	14	15	16	17	18	19	20
21	22	23	24	25	26	27	28	29	30
31	32	33	34	35	36	37	38	39	40
41	42	43	44	45	46	47	48	49	50
51	52	53	54	55	56	57	58	59	60
61	62	63	64	65	66	67	68	69	70
71	72	73	74	75	76	77	78	79	80
81	82	83	84	85	86	87	88	89	90
91	92	93	94	95	96	97	98	99	100
101	102	103	104	105	106	107	108	109	110

Fraction/Decimal Concentration Cards

NAME   DATE   TIME

$\dfrac{1}{10}$	$\dfrac{2}{10}$	$\dfrac{3}{10}$	$\dfrac{4}{10}$
$\dfrac{5}{10}$	$\dfrac{6}{10}$	$\dfrac{7}{10}$	$\dfrac{8}{10}$
$\dfrac{9}{10}$	$\dfrac{10}{10}$	$\dfrac{1}{100}$	$\dfrac{2}{100}$
$\dfrac{3}{100}$	$\dfrac{4}{100}$	$\dfrac{5}{100}$	$\dfrac{6}{100}$
$\dfrac{7}{100}$	$\dfrac{8}{100}$	$\dfrac{9}{100}$	$\dfrac{100}{100}$

G32

Fraction/Decimal
Concentration Cards
(continued)

NAME   DATE   TIME

0.1	0.2	0.3	0.4
0.5	0.6	0.7	0.8
0.9	1.0	0.01	0.02
0.03	0.04	0.05	0.06
0.07	0.08	0.09	1.00

Dollar Exchange Game Mat

One-Dollar Bills	Ten-Dollar Bills	Hundred-Dollar Bills

G34

Multiplication
Wrestling Record Sheet

NAME   DATE   TIME

SRB
267

Round 1   Cards: _____ _____ _____ _____

Numbers formed: _____ * _____

Teams: (_____ + _____) * (_____ + _____)

Products: _____ * _____ = _____

_____ * _____ = _____

_____ * _____ = _____

_____ * _____ = _____

Total (add 4 products): _____

Round 2   Cards: _____ _____ _____ _____

Numbers formed: _____ * _____

Teams: (_____ + _____) * (_____ + _____)

Products: _____ * _____ = _____

_____ * _____ = _____

_____ * _____ = _____

_____ * _____ = _____

Total (add 4 products): _____

Round 3   Cards: _____ _____ _____ _____

Numbers formed: _____ * _____

Teams: (_____ + _____) * (_____ + _____)

Products: _____ * _____ = _____

_____ * _____ = _____

_____ * _____ = _____

_____ * _____ = _____

Total (add 4 products): _____

G35

Expanding
Rugs and Fences

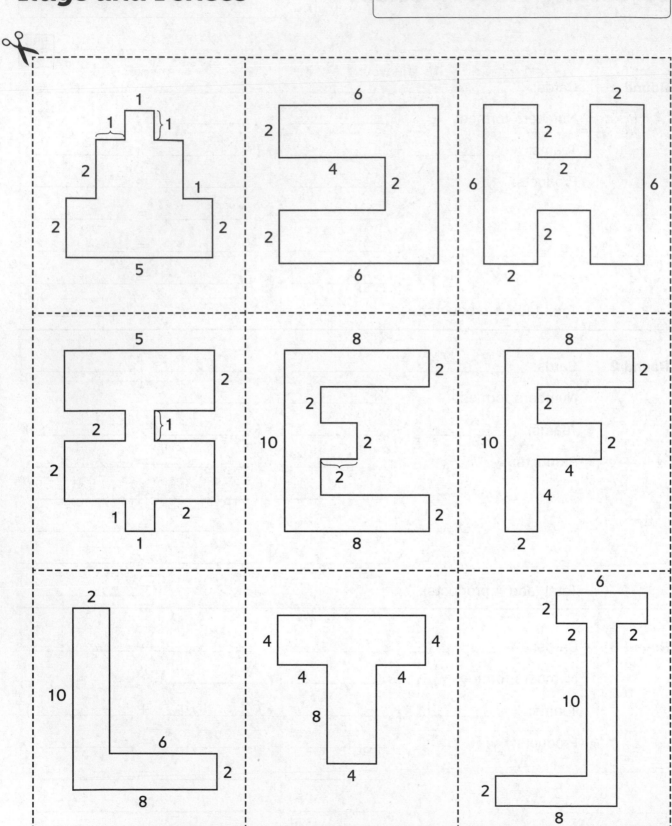

G36

Fishing for Fractions
Record Sheet

SRB 260

NAME   DATE   TIME

Round	Equation with Unknown	Answer
Sample	$\frac{3}{8} + \frac{2}{8} = m$	$\frac{5}{8}$
1		
2		
3		
4		
5		
6		

✂ -

Fishing for Fractions
Record Sheet

SRB 260

NAME   DATE   TIME

Round	Equation with Unknown	Answer
Sample	$\frac{3}{8} + \frac{2}{8} = m$	$\frac{5}{8}$
1		
2		
3		
4		
5		
6		

Divide and Conquer
Fact Triangles

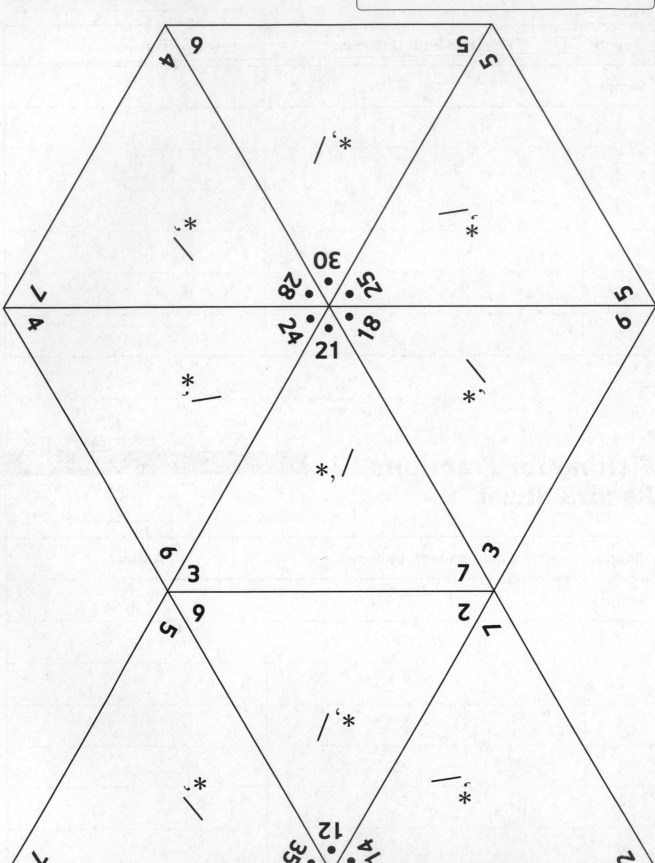

Divide and Conquer
Fact Triangles (continued)

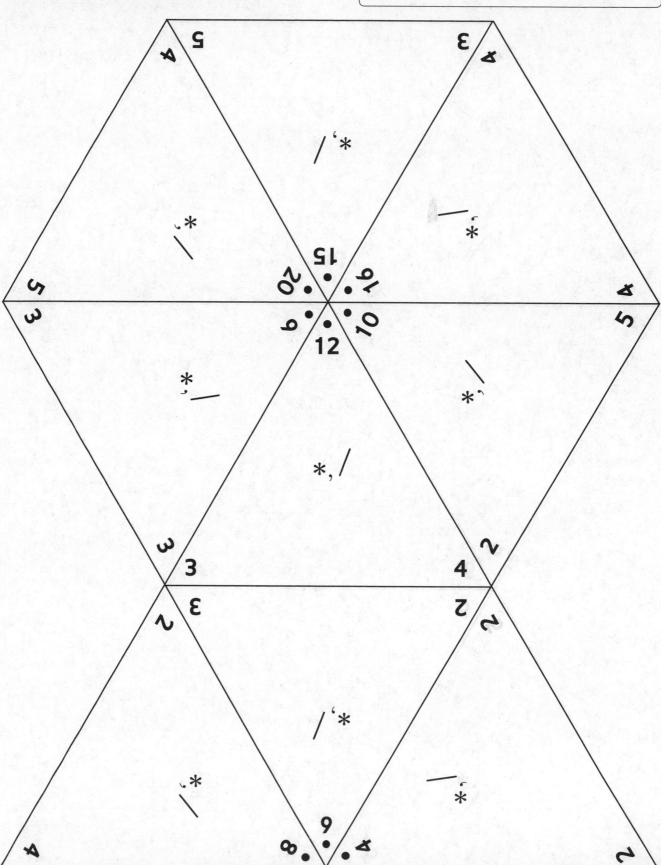

G39

Divide and Conquer
Fact Triangles (continued)

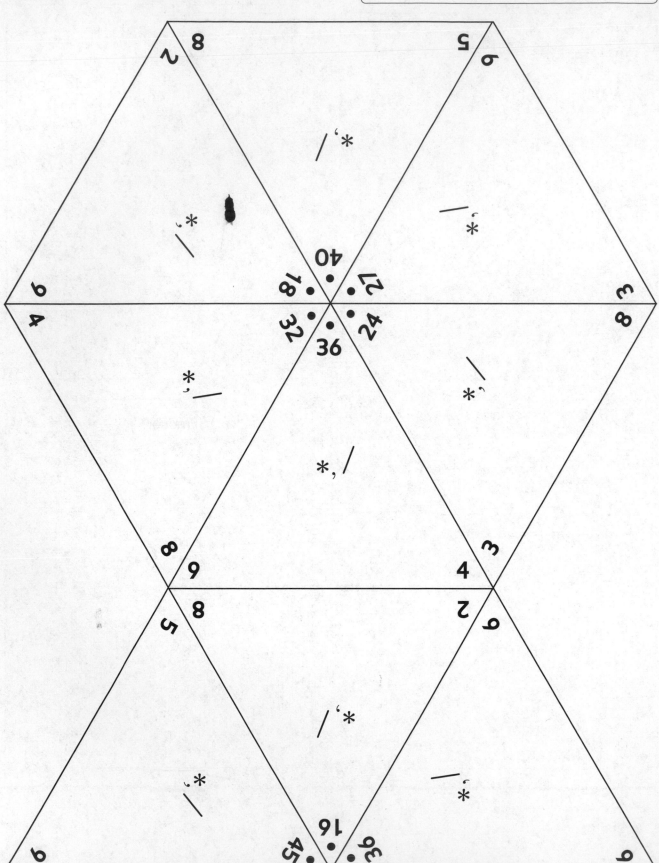

G40

Fishing for Fractions
(Mixed-Number Addition)

NAME   DATE   TIME

Number Card	Fraction Card

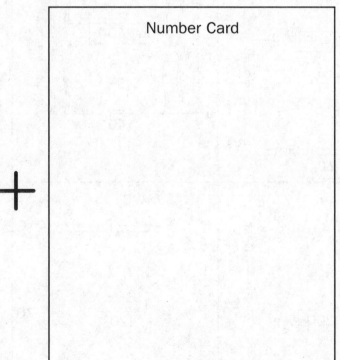

+

Number Card	Fraction Card

Rugs and Fences Missing Side Length Cards

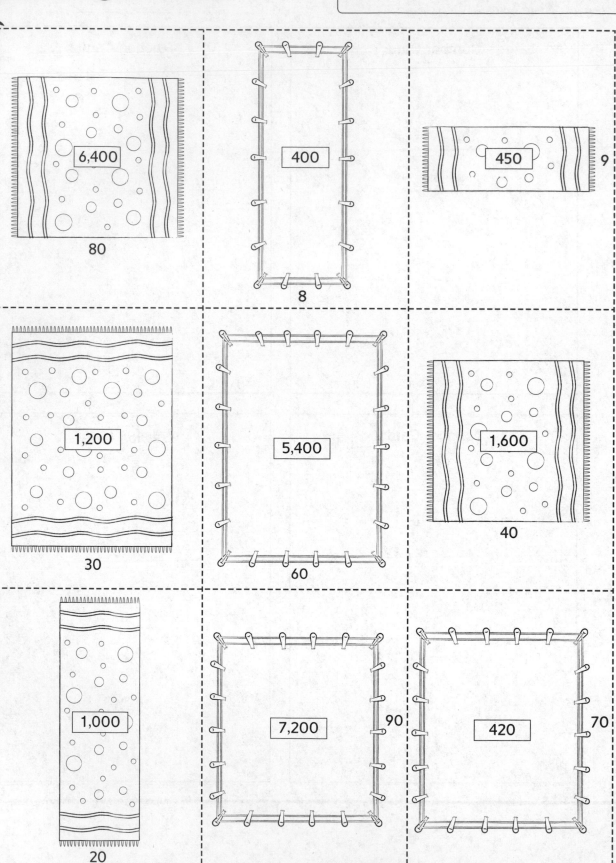

Fishing for Fractions (Mixed-Number Subtraction)

Larger Whole Number Card

Fraction Card

Smaller Whole Number Card

Fraction Card

G43

Beat the Calculator
Gameboard

card 1   *   card 2   O

card 1   O *   card 2

card 1   O *   card 2   O

Division Dash
Record Sheet

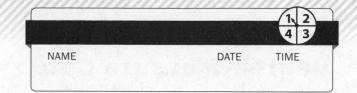

	Division Problem	Quotient	Score
Sample	49 ÷ 4	12 R1	12
1			
2			
3			
4			
5			

✂ -

Division Dash
Record Sheet

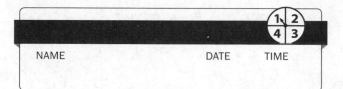

	Division Problem	Quotient	Score
Sample	49 ÷ 4	12 R1	12
1			
2			
3			
4			
5			

Angle Race
Degree-Measure Cards

15°	15°	15°	15°	30°	30°
30°	30°	45°	45°	45°	45°
60°	60°	60°	75°	75°	75°
90°	90°	90°	120°	120°	150°
180°	210°	240°			

Angle Add-Up
Record Sheet

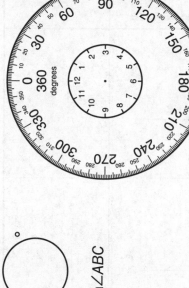

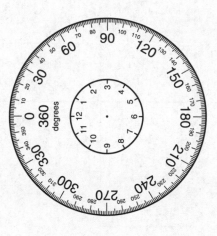

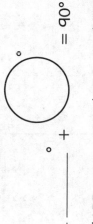

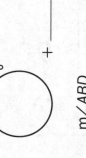

Round 1:

Draw 4 cards.

m∠ABD ____ + ____ m∠DBC = ____ m∠ABC

Round 2:

Draw 2 cards.

m∠ABD ____ + ____ m∠DBC = 90° m∠ABC

Round 3:

Draw 2 cards.

m∠ABD ____ + ____ m∠DBC = 180° m∠ABC

Total points = ____

Angle Tangle
Record Sheet

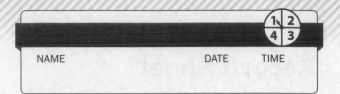

NAME   DATE   TIME

Round	Angle	Estimated Measure	Actual Measure	Score
1		_____°	_____°	
2		_____°	_____°	
3		_____°	_____°	
4		_____°	_____°	
5		_____°	_____°	
			Total Score	

Name That Number
Record Sheet

NAME	DATE	TIME

SRB
268

Round 1

Target Number: _____ My Cards: _____ _____ _____ _____ _____

My best solution (number sentence): _____

Number of cards used: _____

Round 2

Target Number: _____ My Cards: _____ _____ _____ _____ _____

My best solution (number sentence): _____

Number of cards used: _____

✂ --

Name That Number
Record Sheet

NAME	DATE	TIME

SRB
268

Round 1

Target Number: _____ My Cards: _____ _____ _____ _____ _____

My best solution (number sentence): _____

Number of cards used: _____

Round 2

Target Number: _____ My Cards: _____ _____ _____ _____ _____

My best solution (number sentence): _____

Number of cards used: _____